SEMA SIMAI Springer Series

Volume 6

More information about this series at http://www.springer.com/series/10532

Antonio Sellitto • Vito Antonio Cimmelli •
David Jou

Mesoscopic Theories of Heat Transport in Nanosystems

 Springer

Antonio Sellitto
Department of Mathematics
Computer Science and Economics
Università della Basilicata
Potenza, Italy

and

Department of Industrial Engineering
Univesità degli Studi di Salerno
Salerno, Italy

Vito Antonio Cimmelli
Department of Mathematics
Computer Science and Economics
Università della Basilicata
Potenza, Italy

David Jou
Department of Physics
Universitat Autònoma de Barcelona
Bellaterra, Catalonia, Spain

ISSN 2199-3041   ISSN 2199-305X   (electronic)
SEMA SIMAI Springer Series
ISBN 978-3-319-27205-4   ISBN 978-3-319-27206-1   (eBook)
DOI 10.1007/978-3-319-27206-1

Library of Congress Control Number: 2016930309

Springer Cham Heidelberg New York Dordrecht London

Springer International Publishing AG Switzerland is part of Springer Science+Business Media
(www.springer.com)

The Authors at *12th Joint European Thermodynamics Conference* in Brescia, Italy.
From *left* to *right*: Antonio Sellitto, Vito Antonio Cimmelli, David Jou

Preface

Heat transport has been an essential topic in the foundations of thermodynamics since the beginnings of the nineteenth century. Fourier's mathematical description of heat transport served as a stimulus for mathematics (Fourier transforms arose because of the need to solve the classical heat equation following from the Fourier law), for physics (it provided a model and source of inspiration for the mathematical description of other transport phenomena, such as Fick's diffusion law and Ohm's electric transport law), and for natural philosophy (it provided a mathematical framework for an irreversible phenomenon, in contrast to Newton's mathematical framework for reversible mechanics). A century and a half before Fourier, Newton had formulated his heat transfer law for the cooling of bodies, or for heat exchange, but without providing a sufficiently wide mathematical framework. Half a century after Fourier, the Stefan-Boltzmann law provided a mathematical basis for radiative heat exchange. Since then, heat transport analysis in its different forms (conduction, convection and radiation) has been a classical topic in physics, engineering, and natural sciences of life and Earth.

Since the closing decades of the twentieth century, heat transport has been experiencing a true revolution, enlarging its domain of applicability and finding new regimes and phenomenologies where Fourier's theory is no longer applicable. This new epoch in heat transfer has been stimulated by miniaturization, but it was preceded, in some ways, by the earlier technological frontier of aerospace engineering due to the need of studying heat transfer and cooling of bodies in rarefied gases.

In both cases, the new aspects have arisen in association with the relation between the heat carriers' mean-free path ℓ and a relevant characteristic size of the system L, expressed by the Knudsen number $\text{Kn} = \ell/L$. Fourier law (as well as classical hydrodynamics and continuous theory in general) are valid in the limit of a very small Knudsen number, i.e., when $\ell/L \ll 1$. Indeed, Kn may increase both because of an increase in ℓ (as in rarefied gases and aerospace engineering), and because of a reduction in L (as in miniaturization technologies).

The recent progress in nanotechnology and its huge economic impact have brought Kn for heat and electric transport to values at which neither Fourier law, nor classical continuum thermodynamics are strictly applicable. Ways to extend the domain of application of thermodynamics to nanosystems are being sought from the perspective of heat transport, in the same way as Carnot researches on thermodynamics were stimulated by the industrial revolution brought up by heat engines. Currently, research is being stimulated by three industrial revolutions: miniaturization (the need to refrigerate supercomputers, where much heat is dissipated in a tiny space), energy management (the need to develop sustainable energy sources such as photovoltaic and thermoelectric ones, which may be more efficient at the nanoscale than in bulk systems), and material sciences (where nanostructures such as, for instance, superlattices, carbon nanotubes, graphene, silicon nanowires, and porous materials may be decisive for the control of heat transport for insulation or refrigeration, or for delicate phonon control in heat rectification and thermal transistors).

Thus, these are exciting and challenging times for heat transfer and thermodynamics. Much of this impetus is coming from microscopic approaches, either from several versions of kinetic theory or fluctuation–dissipation theorems, or from detailed computer simulations. These approaches allow for detailed understanding and description. However, this should not make us forget the practical usefulness and the conceptual challenge of mesoscopic approaches starting from the macroscopic perspective and deepening into more detailed and accurate descriptions of physical systems.

This is the principal aim of the present book: how to formulate, from a mesoscopic perspective, generalized transport laws able to keep pace with current microscopic research, and to cope efficiently with new applications. The equations presented here are compatible with generalized formulations of nonequilibrium thermodynamics beyond local equilibrium. However, we try to emphasize the transport equations by themselves. When the mean-free path and relaxation times are negligible with respect to the characteristic size of the system and the rates of phenomena, respectively, these equations reduce to Fourier law, but in other situations they describe other physical features beyond it: heat waves, ballistic transport, and phonon hydrodynamics, for instance. We have tried to connect, as far as is possible our results with the results of microscopic theories. This is the reason, for instance, why we use the concept of phonon hydrodynamics as a denomination for a given regime of heat flow where the equations for the heat flux have a form analogous to the hydrodynamic equations for the velocity field. In fact, this form is derived here from the mesoscopic equations in some regimes, but without making explicit reference to the physical nature of heat carriers (they could be phonons, or electrons, or holes), but since the phenomenology known to date has been explored microscopically, we have retained the name "phonon hydrodynamics" to enhance the connection between the mesoscopic and the microscopic approach.

Potenza, Italy   Antonio Sellitto
Potenza, Italy   Vito Antonio Cimmelli
Bellaterra, Catalonia   David Jou

Acknowledgements

Antonio Sellitto acknowledges the University of Basilicata for funding the research project *Modeling heat and electric transport in nanosystems in the presence of memory, nonlocal and nonlinear effects*, and the Italian National Group of Mathematical Physics (GNFM-INdAM) for financial support.

Vito Antonio Cimmelli acknowledges the financial support of the Italian National Group of Mathematical Physics (GNFM-INdAM). Thanks are also due to the University of Basilicata for financial support and for funding the research subject in Mathematical Physics *Equazioni costitutive per la conduzione del calore nei nanosistemi*.

David Jou acknowledges the financial support provided by the *Dirección General de Investigación* of the Spanish Ministry of Science and Innovation under grant FIS No. 2009-13370-C0201, the *Consolider Project NanoTherm* (grant CSD-2010-00044), the Ministry of Economy and Competitiveness under grant FIS2012-32099, and the *Direcció General de Recerca* of the Generalitat of Catalonia under grant No. 2009-SGR-00164.

Acknowledgements

About This Book

In Chap. 1 we provide a sketch of Extended Irreversible Thermodynamics (EIT) which represents the general framework within which the generalized heat transport equations are built up.

In EIT one starts with an entropy and an entropy flux which depend on the fluxes, as well as on the classical variables. Not too far from equilibrium the entropy is the local equilibrium entropy minus some quadratic expressions in the fluxes, and the entropy flux is the classical entropy flux plus a correction proportional to the product of the flux of the flux times the flux itself. From here one is led for the entropy production to a quadratic form which is a sum of products of fluxes times generalized thermodynamic forces, and it is required that the constitutive equations are submitted to the restriction that the entropy production is always positive definite. The main differences with respect to Classical Irreversible Thermodynamics (CIT) are: (1) The thermodynamic forces are more general than their local equilibrium counterparts, and in them appear the time derivatives and the gradients of the fluxes, as a consequence of using an entropy and entropy flux which depend on the fluxes. (2) Since the fluxes are considered as independent variables, the constitutive equations do not aim to express the fluxes in terms of thermodynamic forces, but to express the evolution equations for the fluxes, i.e., to give the time derivative of the fluxes in terms of classical thermodynamic forces, the fluxes, and the gradients of the fluxes.

After the evolution equations for the fluxes have been formulated, one is able to give a physical interpretation to the several coefficients appearing in such equations (and related to the coefficients of the nonequilibrium contributions to the entropy and the entropy flux) in terms of the classical transport coefficients, the relaxation times of the fluxes, the correlation lengths of the fluxes, and so on. As a simple illustration we mention the situation related to the so-called Guyer-Krumhansl equation for heat transfer in rigid conductors, which takes into account relaxation effects and nonlocal effects. The corresponding entropy depends on the internal energy as well as on the heat flux, and the corresponding entropy flux depends on the gradient of the heat flux, too. When the mean-free path is negligible, the Guyer-Krumhansl equation reduces to Cattaneo's equation.

Finally, the absolute temperature is given by the reciprocal of the derivative of the entropy with respect to the internal energy (at constant values of the other extensive variables). When the extended entropy is used instead of the local-equilibrium entropy, the resulting absolute temperature differs from the local-equilibrium temperature, and depends on the fluxes.

In Chap. 2 we provide an overview of the different generalized heat transport equations which are able to take into account nonlinear, nonlocal and memory effects. All these models stem from a precise point of view on the problem of heat conduction. On this subject, the celebrated paper by Cattaneo, in which it is observed that the classical heat equation leads to the "paradox" of infinite speeds of propagation of the thermal disturbances, stimulated research in various directions. Hence, new theories have been formulated and new schools of thermodynamics have been founded in recent decades. Although Cattaneo was not a thermodynamicist, he contributed much to the development of modern thermodynamics because his paper stimulated the formulation of new theories.

All the presented models are able to reproduce some of the main properties of heat conduction at low temperatures, but none of them leads to an exhaustive description which is free of technical problems. In particular, in the extended thermodynamic theories, in which the fluxes are included in the state space, the higher order fluxes are necessary in modeling high-frequency processes and the memory effects are modeled by the hierarchical system of equations which is a consequence of kinetic theory, the main problems are due to the rapidly increasing number of unknown quantities. The most important of these is the determination of the appropriate number of equations, i.e., of the step at which the hierarchical system should be truncated. A further problem is related to the Guyer-Krumhansl equation, which plays a fundamental role in describing nonlocal and relaxation effects. Such an equation is parabolic, so that finite speeds of propagation, in a strict mathematical sense, cannot be expected. However finite speeds of propagation can be obtained in the generalized sense as clarified by G. Fichera and W.A. Day. These authors observed that correct estimation of the speed of propagation of the disturbances requires that the order of magnitude of the solution of the system of balance equations is compared with that of the error affecting the available experimental data which the theory aims to reproduce. In particular, if after a finite interval of time from the beginning of a physical process the solution of the corresponding system of balance laws is greater than the experimental error only in a compact domain, then we can say that such a solution is experimentally zero outside this domain. As a consequence, it has propagated with finite speed because not all the points of the space have been reached in a finite time.

In Chap. 3 we develop a mesoscopic description of boundary effects and effective thermal conductivity in nanosystems by using phonon hydrodynamics. Nanometer-sized devices are of considerable current interest in micro/nanoelectronics, since currently miniaturization is being successfully applied in several technologies, including optics, catalysis, and ceramics.

Indeed, the rise and the development of nanotechnology requires increasing efforts to better understand the thermal transport properties of nanodevices, as their performance and reliability are much influenced by memory, nonlocal and nonlinear effects. The macroscopic derivation of generalized transport equations represents a very important step in achieving this task. This problem may be tackled by different approaches, the most frequently pursued of which is the microscopic one, based upon the Boltzmann equation. An alternative approach is that based on so-called phonon hydrodynamics, which regards the entire set of heat carriers as a fluid whose hydrodynamic-like equations describe the heat transport. This mesoscopic approach allows for a fast quantitative approximate estimation of the thermal properties and may be a useful complement to microscopic theories in order to select the most promising features of nanosystems. In the linear regime the phonon hydrodynamics rests on the Guyer-Krumhansl transport equation. However, it is worth observing that in the original proposal of Guyer and Krumhansl a boundary relaxation time was added to the usual relaxation time due to resistive mechanisms by use of the Matthiessen rule. Once the combined resistive-boundary collision time has been obtained, the thermal conductivity (depending on the size of the system through the boundary relaxation time) can be calculated and used in the Guyer-Krumhansl transport equation. Here, instead, we use a different method, which consists in including the boundary collision time not in the differential equation, but in suitable boundary constitutive equations. Particular attention is paid to the modeling of the constitutive equations for a slip heat flux along the walls.

In Chap. 4 we describe in some detail the application of the model of phonon hydrodynamics to nanoporous materials.

In fact, in addition to referring to small systems whose characteristic length is of the order of nanometers, the name "nanosystem" can also refer to those systems characterized by an internal nanostructure that gives them some special mechanical, thermal, electrical and optical properties. Such structures may be nanopores, or nanoparticles, or several parallel and very thin layers (graded materials). By regulating the main features of such an internal structure, one may control the transport properties of those systems.

When the characteristic size of the system is large as compared with the mean-free path of the heat carriers, the thermal properties can be derived in the framework of the classical Fourier's theory. It is more difficult to analyze such properties when the aforementioned features are comparable to (or smaller than) the mean-free path of heat carriers, in which case several new problems emerge.

We also briefly furnish in less detail, some results from other theories for the thermal conductivity both of nanocomposites and superlattices, and of nanofluids, which are the basis for many outstanding applications.

In Chap. 5 we illustrate some peculiar phenomena which can be identified when weakly nonlocal and nonlinear heat transport equations are used. As stated above, the thermal behavior of systems whose characteristic length is of the order of a few nanometers is strongly influenced by memory, nonlocal, and nonlinear effects. In one-dimensional steady-state situations, and when modeling the heat transport along nanowires or thin layers, some of these effects may be incorporated into a size-dependent effective thermal conductivity and a Fourier-type equation may still be used with an effective value of the thermal conductivity. However, in fast perturbations, or under strong heat gradients, an effective thermal conductivity is insufficient to overcome the different problems related to the Fourier law and therefore, in modeling heat conduction, it is necessary to go beyond such an equation.

In addition to presenting some results that are important from the theoretical point of view, such as the existence of flux limiters, we also illustrate several applications, including thermal rectification in troncoconical nanowires and axial heat propagation in thin layers and graphene sheets.

Also, based on the methods of classical hydrodynamics, a mathematical procedure is developed which is proven to be useful in studying the stability of the heat flow in nanowires.

In Chap. 6 we study how to obtain enhanced versions of classical thermoelectric equations, which have meaningful consequences for some well-known classical theoretical results and in practical applications, e.g., for the efficiency of thermoelectric energy generators. In fact, thermoelectric devices offer an attractive source of energy, since they do not have moving parts, do not create pollution, and do not emit noise. Nanomaterials provide an interesting avenue by which to obtain better performing thermoelectric devices, for example, by making nanocomposites, adding nanoparticles to a bulk material, or using one-dimensional nanostructures. For the sake of illustration, we consider a cylindrical thermoelectric nanodevice whose longitudinal length L is much larger than the characteristic size of the transverse section. One may represent it as a one-dimensional system and consider only one Cartesian component, namely, the longitudinal one. In steady states we assume that the hot side of this system is kept at temperature T^h and the cold side at temperature T^c. Moreover, we suppose that an electric current and a quantity of heat per unit time enter uniformly into the hot side of the device and flow through it.

In this chapter we further point out nonlocal and/or nonlinear breaking of Onsager symmetry, which implies modifications of the expression for the maximum efficiency in terms of the transport coefficients and the temperature.

All the results derived herein rest on the basic assumption that in thermoelectric materials the local heat flux has two different contributions, namely, the phonon partial heat flux and the electron partial heat flux. In the simplest situation, the heat carriers (phonons and electrons) may be supposed to have the same temperature. However, in more complex situations, such as in the presence of "hot electrons", or when a laser heat pulse hits a material surface, the phonon and electron temperatures may be different. Thus, in this chapter, the classical results of thermoelectricity are revisited under the hypothesis that phonons and electrons have different temperatures.

Like Lennon and McCartney, we can say that this book has been written "With a Little Help from Our Friends". In fact, in aiming at a deeper understanding of the topics developed in the present book, we had the great advantage of being able to discuss them with several of our colleagues and friends from around the world. In particular, we acknowledge our fruitful discussions with J. Casas-Vázquez, F.X. Alvarez, J. Camacho, J. Bafaluy and V. Méndez (Barcelona, Spain), Z.-Y. Guo,

M. Wang, Y. Dong and Y. Guo (Beijing, China), W. Muschik (Berlin, Germany), T. Ruggeri (Bologna, Italy), P. Ván (Budapest, Hungary), R. Luzzi (Campinas, Brazil), J. Fort (Girona, Spain), G. Lebon (Liege, Belgium), F. Oliveri, L. Restuccia and P. Rogolino (Messina, Italy), M. Grmela (Montreal, Canada), F. Vázquez (Morelos, Mexico), M.S. Mongioví, M. Sciacca and L. Saluto (Palermo, Italy), V. Triani and I. Carlomagno (Potenza, Italy), K. Frischmuth (Rostock, Germany), and W. Kosiński (Warsaw, Poland, who passed away in 2014).

We also like to note that during the proof correction of this book it has appeared the review paper

Guo, Y., Wang, M.: Phonon hydrodynamics and its applications in nanoscale heat transport, Physics Reports, 595, 1-44 (2015)

It provides interesting results related to the topics discussed in this book.

Potenza, Italy   Antonio Sellitto
Potenza, Italy   Vito Antonio Cimmelli
Bellaterra, Catalonia, Spain   David Jou

Contents

with $s_{eq}(u)$ representing the entropy at the equilibrium (that is, in the absence of $\mathbf{q}$), and u being the internal energy per unit volume, which is ruled by the following evolution equation

$$\dot{u} + \nabla \cdot \mathbf{q} = 0 \qquad (1.8)$$

in the absence of heat source. Although Eq. (1.7) is not the very general constitutive equation one may introduce, we explicitly note that it is in accordance with the general theorems of representation of isotropic scalar functions depending on scalars and vectors [84].

On the other hand, the local balance of entropy reads

$$\dot{s} + \nabla \cdot \mathbf{J}^s = \sigma^s \qquad (1.9)$$

with $\mathbf{J}^s$ as the entropy flux, and σ^s as the entropy production per unit volume. Once the usual assumptions

$$\frac{\partial s}{\partial u} = \frac{1}{T} \qquad (1.10\text{a})$$

$$\mathbf{J}^s = \frac{\mathbf{q}}{T} \qquad (1.10\text{b})$$

are made, the coupling of Eqs. (1.6)–(1.9) yields

$$\sigma^s = \mathbf{q} \cdot \left[\nabla T^{-1} + s_1 \dot{\mathbf{q}} \right] . \qquad (1.11)$$

Second law of thermodynamics dictates that the left-hand side of Eq. (1.11) has to be positive whatever the thermodynamic process is [33]. Since the right-hand side of that equation is in the form of product between the thermodynamic flux $\mathbf{q}$ and the thermodynamic force $\nabla T^{-1} + s_1 \dot{\mathbf{q}}$, the constrain $\sigma^s \geq 0$ can be fulfilled, for example, if [36]

$$\nabla T^{-1} + s_1 \dot{\mathbf{q}} = L_1 \mathbf{q} \qquad (1.12)$$

where the phenomenological coefficient L_1 may depend both on u, and on $\mathbf{q}$ in principle. In particular, the identifications

$$L_1 = \frac{1}{\lambda T^2}$$
$$s_1 = -\frac{\tau_R}{\lambda T^2}$$

allow Eq. (1.12) to reduce to the MCV equation (1.6), proving so its physical consistency. It is worth observing that the constitutive equation (1.7) in this case takes the form

$$s = s\left(u, \mathbf{q}\right) = s_{\text{eq}}\left(u\right) - \frac{\tau_R}{2\lambda T^2}\mathbf{q} \cdot \mathbf{q} \qquad (1.13)$$

which also guarantees that the principle of maximum entropy at the equilibrium [33] is always satisfied.

A similar procedure can be applied if $\mathbf{J}$ is a different flux, as for instance, the flux of mass, or the flux of momentum in classical continuum mechanics, one of the partial fluxes of mass or momentum in theory of mixtures [36], the flux of electric charge in thermoelectricity [64]. In all the mentioned cases, the main task is to assign constitutive equations for the entropy density and for the entropy flux which are physically sound and manageable from the mathematical point of view.

1.2   Second-Order Nonlocal Effects

When the ratio between the mean-free path ℓ of heat carriers and the characteristic length L of the system (i.e., the so-called Knudsen number Kn $= \ell/L$) becomes comparable to (or higher than 1), the heat transport is no longer diffusive, but ballistic. Moreover, when the mean-free path between successive collisions becomes large, there will be a direct connection among nonadjacent regions of the system with very different values of the temperature. In this situation it may be asked whether laws like Eqs. (1.2) and (1.6) remain applicable, or not [66].

Since nonlocal effects are especially important to describe the transition from diffusive to ballistic regime, to account for them in the framework of EIT it is possible to introduce a new extra variable, represented by a second-order tensor $\mathbf{Q}$, and write the balance of the heat flux in the following general form [56]

$$\tau_1 \dot{\mathbf{q}} + \mathbf{q} = -\lambda \nabla T + \nabla \cdot \mathbf{Q} \qquad (1.14)$$

where τ_1 is still a relaxation time. The tensor $\mathbf{Q}$, assumed to be symmetric, may be split in the usual form $\mathbf{Q} = Q\mathbf{I} + \mathbf{Q}_s$, the scalar Q being one-third of the trace of $\mathbf{Q}$, and $\mathbf{Q}_s$ being the deviatoric part of $\mathbf{Q}$. In the relaxation time approximation [67], the evolution equations for Q and $\mathbf{Q}_s$ may be written as

$$\tau_0 \dot{Q} + Q = \gamma_0 \nabla \cdot \mathbf{q} \qquad (1.15a)$$

$$\tau_2 \dot{\mathbf{Q}}_s + \mathbf{Q}_s = 2\gamma_2 \left[(\nabla \mathbf{q})^0\right]_s \qquad (1.15b)$$

where $\left[(\nabla \mathbf{q})^0\right]_s$ denotes the symmetric and traceless part of $\nabla \mathbf{q}$. Assuming that the relaxation times τ_0 and τ_2 are negligibly small, and considering only regular solutions for which the time derivatives appearing at the left-hand side of Eqs. (1.15a) and (1.15b) do not diverge, and substituting these equations into Eq. (1.14), one obtains for the evolution of the heat flux

$$\tau_1 \dot{\mathbf{q}} + \mathbf{q} = -\lambda \nabla T + \gamma_2 \nabla^2 \mathbf{q} + \left(\gamma_0 + \frac{1}{3}\gamma_2 \right) \nabla \nabla \cdot \mathbf{q}. \tag{1.16}$$

Equation (1.16) is comparable with that obtained by Guyer and Krumhansl from phonon kinetic theory [47, 48], namely,

$$\tau_R \dot{\mathbf{q}} + \mathbf{q} = -\lambda \nabla T + \ell_p^2 \left(\nabla^2 \mathbf{q} + 2\nabla \nabla \cdot \mathbf{q} \right) \tag{1.17}$$

wherein ℓ_p is the phonon mean-free path. It is easy to recognize, in fact, that the general heat-transport equation (1.16) reduces to the Eq. (1.17) under the following identifications

$$\tau_1 = \tau_R, \ \gamma_0 = \frac{5}{3}\ell_p^2, \ \gamma_2 = \ell_p^2. \tag{1.18}$$

It is worth observing that in Refs. [47, 48] Guyer and Krumhansl deal with resistive and normal collisions, and relate ℓ_p to both processes. Here we are mainly interested in the mathematical aspects of Eq. (1.17), rather than in its microscopic derivation.

The thermodynamics underlying the transport equation (1.16) is easily derived by following the same procedure as in Sect. 1.1, i.e., by introducing extended constitutive equations for the entropy density and for the entropy flux of the form, respectively [53, 58],

$$s = s_{eq}(u) - \frac{\tau_1}{2\lambda T^2} \mathbf{q} \cdot \mathbf{q} - \frac{\tau_2}{4\lambda T^2 \gamma_2} \mathbf{Q}_s : \mathbf{Q}_s - \frac{\tau_0}{2\lambda T^2 \gamma_0} Q^2 \tag{1.19a}$$

$$\mathbf{J}^s = \frac{\mathbf{q}}{T}\left(1 + \frac{Q}{\lambda T} \right) + \frac{\mathbf{Q}_s * \mathbf{q}}{\lambda T^2} \tag{1.19b}$$

where the symbol : denotes the complete contraction of the corresponding tensors giving a scalar as result.

Note, incidentally, that Eq. (1.19a) generalizes Eq. (1.13) and is still in agreement both with theorems of representation for scalar functions depending on scalars, vectors, and tensors [84], and with the principle of maximum entropy at the equilibrium [33]. Equation (1.19b), instead, agrees with the general form of the entropy current derived by Verhas [93], according to which the entropy flux is the sum of all fluxes each of them multiplied by the suitable intensive quantities appearing in the entropy expression.

1.3 Higher-Order Fluxes and Nonlinear Hierarchy of Transport Equations: Ballistic Heat Transport

So far, only the heat flux $\mathbf{q}$ and the flux of the heat flux $\mathbf{Q}$ have been assumed as nonequilibrium variables. Indeed, the kinetic theory points out that the relaxation times of the higher-order fluxes are not always shorter than the collision time. The only use of the first-order fluxes as independent variables is not satisfactory to describe high-frequency processes, because when the frequency becomes comparable to the inverse of the relaxation time of the first-order flux, all the higher-order fluxes will also behave like independent variables and must be incorporated in the formalism [56, 58, 81].

In EIT it is possible to incorporate higher-order variables, each one being defined as the flux of the preceding one, and denoted as $\mathbf{J}_1, \mathbf{J}_2, \ldots, \mathbf{J}_n$, where $\mathbf{J}_k$ with $k = 1 \ldots n$, is a tensor of the order k and stands for the flux of the element of the hierarchy of order $\mathbf{J}_{k-1}$.

Up to the n-th order moment, a convenient constitutive equation for the volumetric entropy takes the form

$$s = s(u, \mathbf{J}_1, \ldots, \mathbf{J}_n) = s_{\text{eq}}(u) + \frac{\alpha_1}{2} \mathbf{J}_1 : \mathbf{J}_1 + \ldots + \frac{\alpha_n}{2} \mathbf{J}_n : \mathbf{J}_n \tag{1.20}$$

while the entropy flux can be written as

$$\mathbf{J}^s = \frac{\mathbf{J}_1}{T} + \beta_1 \mathbf{J}_2 * \mathbf{J}_1 + \ldots + \beta_n \mathbf{J}_n \odot \mathbf{J}_{n-1} \tag{1.21}$$

with $\alpha_1 \ldots \alpha_n$ and $\beta_1 \ldots \beta_n$ being suitable functions of material coefficients which are supposed to depend on u, and the notation $\mathbf{J}_k \odot \mathbf{J}_{k-1}$ means the contraction over the last $k-1$ indices of $\mathbf{J}_k$, giving a first-order tensor, i.e., a vector, as result. It is worth observing that in the previous expressions of the entropy and of the entropy flux, cubic coupled contributions have not been taken into account for the sake of a formal simplicity.

The corresponding evolution equations, compatible with a positive entropy production, are written in the following hierarchy [44]

$$\tau_1 \dot{\mathbf{J}}_1 + \mathbf{J}_1 = -\lambda \nabla T + \frac{\beta_1 \tau_1}{\alpha_1} \nabla \cdot \mathbf{J}_2$$

$$\vdots \tag{1.22}$$

$$\tau_n \dot{\mathbf{J}}_n + \mathbf{J}_n = \frac{\beta_n \tau_n}{\alpha_n} \nabla \cdot \mathbf{J}_{n+1} + \frac{\beta_{n-1} \tau_n}{\alpha_n} \nabla \mathbf{J}_{n-1}.$$

In the system of Eq. (1.22) the compatibility with the general balance law for the fluxes

$$\dot{\mathbf{J}}_k + \nabla \cdot \mathbf{J}_{k+1} = \mathbf{P}_k \tag{1.23}$$

where $k = 2 \ldots n$, and $\mathbf{P}_k$ denotes the production of $\mathbf{J}_k$, implies

$$\beta_k = -\alpha_k, \quad \mathbf{P}_k = -\frac{\mathbf{J}_k}{\tau_k} + \frac{\beta_{k-1}}{\beta_k} \nabla \mathbf{J}_{k-1}. \tag{1.24}$$

This hierarchy of equations may be used to describe the transition from diffusive to ballistic regime [1, 2, 55, 70]. In fact, it leads to a continued fraction expansion [37, 50] for the thermal conductivity in terms of the Knudsen number as

$$\lambda_{\text{eff}}(T, \text{Kn}) = \frac{\lambda(T)}{1 + \cfrac{(K\ell_1)^2}{1 + \cfrac{(K\ell_2)^2}{1 + \cfrac{(K\ell_3)^2}{1 + \ldots}}}} \tag{1.25}$$

wherein $K = 2\pi/L$, and ℓ_i $(i = 1, \ldots n)$ are the effective mean-free paths related to the flux of order i. If in Eq. (1.25) all the coefficient ℓ_i are equal, namely, $\ell_i = \ell_p/2$, it follows that the effective thermal conductivity is [1, 2]

$$\lambda_{\text{eff}} = \frac{\lambda}{2\pi^2 \, \text{Kn}^2} \left(\sqrt{1 + 4\pi^2 \, \text{Kn}^2} - 1 \right). \tag{1.26}$$

This expression describes a strong reduction of the effective thermal conductivity for increasing values of Kn. When $\text{Kn} = 0$, Eq. (1.26) yields $\lambda_{\text{eff}} \equiv \lambda$, whereas for large Kn values it turns out

$$\lambda_{\text{eff}} = \frac{\lambda}{\pi \, \text{Kn}}$$

and leads to the following expression for the modulus of the local heat flux

$$q \approx \frac{c_v \overline{v}}{3\pi} \Delta T$$

were c_v is the specific heat per unit volume, and $\overline{v}$ is the average phonon speed. This behavior, which is independent of the size of the system, is typical of ballistic regime. In qualitative terms, Eq. (1.26) provides a reasonable description of experimental results [1].

Another possibility for Eq. (1.25) would be to assume that the mean-free path ℓ_n, associated with the heat flux of order n, is given as

$$\ell_n^2 = \ell_p^2 \frac{(n+1)^2}{\left[4(n+1)^2 - 1 \right]}$$

as it is usual in phonon's kinetic theory [43]. In this case the asymptotic form of Eq. (1.25) is [69]

$$\lambda_{\mathrm{eff}} = \frac{3\lambda}{4\pi^2 \, \mathrm{Kn}^2} \left[\frac{2\pi \, \mathrm{Kn}}{\arctan{(2\pi \, \mathrm{Kn})}} - 1 \right] \tag{1.27}$$

which, however, behaves in a similar way as Eq. (1.26). The expressions in Eqs. (1.26) and (1.27) for the effective thermal conductivity are more sophisticated and precise than the usual assumption

$$\lambda_{\mathrm{eff}} = \frac{\lambda}{1 + f(T) \, \mathrm{Kn}} \tag{1.28}$$

where $f(T)$ is nondimensional function depending on temperature and on the form of the cross section of the system. Referring the readers to Chap. 3 for more comments about that equation, here we only observe that Eqs. (1.26)–(1.28) essentially have the same qualitative behavior both for low, and for very high Knudsen numbers.

Several mathematical aspects of Eq. (1.22) and related ones have been dealt with in Ref. [29]. The idea of a linear hierarchy of evolution equations presented above may be generalized by taking into account that, in the presence of a temperature gradient, the fluxes of order m, $m + 1$, and $m - 1$ may be coupled not only by the wave-vector $\mathbf{k}$, but also by the temperature gradient itself. As a consequence the hierarchy of Eq. (1.22) may be generalized as follows [56, 58]

$$\tau_m \dot{\mathbf{J}}_m + \mathbf{J}_m = \lambda_m \nabla \mathbf{J}_{m-1} + \delta_m \nabla \cdot \mathbf{J}_{m+1} + \Lambda_m \nabla T \otimes \mathbf{J}_{m-1} + \Delta_m \mathbf{J}_{m+1} \odot \nabla T \tag{1.29}$$

where δ_m, Λ_m and Δ_m are still material coefficients, and the symbol $\otimes$ indicates the standard tensor product between ∇T and $\mathbf{J}_{m-1}$ turning out a tensor of order m. Shifting to the Fourier-Laplace transform, with the assumptions

$$\mathbf{B}_m = \Delta_m \nabla T + i\delta_m \mathbf{k}, \quad \mathbf{C}_m = \Lambda_m \nabla T + i\lambda_m \mathbf{k}, \tag{1.30}$$

Equation (1.29) may be easily rearranged as

$$\tau_m \dot{\mathbf{J}}_m + \mathbf{J}_m = \mathbf{J}_{m+1} \odot \mathbf{B}_m + \mathbf{C}_m \otimes \mathbf{J}_{m-1}. \tag{1.31}$$

This equation is important in the study of high-frequency processes in systems with heat conductivity depending on the frequency of the thermal waves as well as on the wave vector and on the temperature gradient. In particular, in steady states and for $\Delta_m = \Lambda_m = \ell_p/T$, it describes a flux-limited behavior of the form

$$\mathbf{q} = -\frac{\lambda \nabla T}{\dfrac{1}{2} + \sqrt{\dfrac{1}{4} + \ell_p^2 (\nabla \ln T)^2}}. \tag{1.32}$$

Equation (1.32) yields the classical FL when the term $\nabla \ln T$ is small, whereas when $\nabla \ln T$ gets large, it leads to $q = \lambda T / \ell_p \approx (1/3) \, c_v \bar{v} T$ which is the saturation value for the modulus of the local heat flux.

1.4 Nonequilibrium Temperature and Heat Transport

Temperature represents a basic concept in thermodynamics and it is the topic of several discussions since its definition in nonequilibrium situations is not yet fully understood [35, 56–58, 64, 68]. In fact, the local-equilibrium temperature loses its validity in situations wherein the deviation from equilibrium ensemble is not negligible, as for example the heat propagation in nanosystems. Therefore, for a better understanding of temperature in nonequilibrium states, the exploration of the consequences of the introduction of different definitions of temperature in heat transport may be very useful.

In linear regime different definitions of temperature are equivalent, and lead to the same conclusions both from the experimental, and from the theoretical point of view.

In nonlinear regime, instead, different temperatures may yield different predictions. This is logical, indeed, because out of equilibrium there is no energy equipartition, and every degree of freedom may be assigned its own temperature. The contribution of the several degrees of freedom to the heat flow may also be different (for instance, its corresponding thermal conductivity may be different). Thus, temperature beyond local equilibrium is more complex than the usual identification of temperature as average kinetic energy of the particles.

In particular, it can be shown that the dispersion relation of heat waves along nanowires (or thin layers) which are not isolated from the environment allows to compare different definitions of nonequilibrium temperature, since thermal waves are predicted to propagate with different phase speed depending on the definition of nonequilibrium temperature being used [57, 58]. From the other hand, in highly nonequilibrium situations, as heat transport in nanosystems experiencing fast processes, the set of independent variables has to be enlarged by including the fluxes of extensive quantities (i.e., internal energy, matter and momentum). They can no longer be neglected and a constitutive description of them in terms of the traditional field variables is insufficient to cope with the complexity of the observed phenomenology. It is worth noticing that also in the kinetic theory of gases, it is usual to select the higher moments of the velocity distribution as independent variables [45, 86].

In the following we give two different definitions of nonequilibrium temperature, each of which may be well placed within the general framework of EIT. Comparison between them is made by analyzing their role in the description of thermal-pulse propagation in nonequilibrium situations.

1.4.1 Dynamical Temperature

Consider a moving gas not in thermodynamic equilibrium and denote by G a certain physical quantity carried by each molecule of the gas, and by $\widehat{G}$ the average of G in a small volume around a point P of the gas. For the sake of simplicity, assume $\widehat{G}$ constant with respect to its spatial variables on each of the planes which are orthogonal to a given direction x, namely, assume $\widehat{G} = \widehat{G}(x, t)$. Finally, explicitly suppose that the thermodynamic state of the gas is not far from the equilibrium [24–28, 30].

Moreover, let dN_c mean the number of molecules per unit of volume passing in the time interval dt across the unit area ω of the plane $x = x_0$, whose velocity is in the interval $I_c = [\widetilde{c}, \widetilde{c} + d\widetilde{c}]$, and denote by $dn_{c\vartheta}$ the fraction of dN_c of molecules moving along a given direction r forming with x a solid angle ϑ. It is easily seen that these molecules are contained in an oblique cylinder having base ω, inclination ϑ and length cdt, so that

$$dn_{c\vartheta} = \frac{\sin\vartheta\,\cos\vartheta}{2} d\vartheta\,dt\,dN_c. \tag{1.33}$$

Focus then the attention on a single molecule running along the positive direction of x with a free path of length ℓ under an angle ϑ with respect to x, and crossing ω at the instant $t' \in [t, t + dt]$ with a velocity whose modulus is in the range I_c. Indicate with $\widehat{G}^+(x, t)$ the quantity $\widehat{G}(x, t)$ referred to this molecule, and with $\widehat{G}^-(x, t)$ the same quantity referred to a molecule running along the negative direction of x.

The net amount of $\widehat{G}$ carried by all the molecules of the type considered is given by the difference

$$\sum{}'\widehat{G}^+(x, t) - \sum{}'\widehat{G}^-(x, t)$$

where the prime indicates that the sum is extended to the molecules of the type considered above, only. To evaluate such a sum we use the classical results

$$\sum{}'\ell \approx \lambda_c dn_{c\vartheta}$$
$$\sum{}'\ell^2 \approx \ell_c^2 dn_{c\vartheta}$$

where λ_c and ℓ_c^2 are the mean-free path and the squared mean-free path of the molecules, respectively [11].

By using the relations above, some lengthy but straightforward calculations allow to prove that [24, 28]

$$\sum{}'\widehat{G}^{+}(x,t) - \sum{}'\widehat{G}^{-}(x,t) =$$

$$\left[2\left(\lambda_c \cos\vartheta\right)\frac{\partial\widehat{G}}{\partial x}\Big|_{(x_0,t')} + \left(\frac{\ell^2\cos\vartheta}{\widetilde{c}}\right)\frac{\partial^2\widehat{G}}{\partial x\partial t}\Big|_{(x_0,t')} + 2\left(\frac{v_c\ell^2\cos\vartheta}{\widetilde{c}}\right)\frac{\partial^2\widehat{G}}{\partial x^2}\Big|_{(x_0,t')}\right.$$

$$\left.+2\left(\frac{v_c\ell^2}{\widetilde{c}^2}\right)\frac{\partial^2\widehat{G}}{\partial x\partial t}\Big|_{(x_0,t')} + 2\left(\frac{\ell^2}{\widetilde{c}^2}\right)\left(\frac{\partial v_c}{\partial t}\right)\frac{\partial\widehat{G}}{\partial x}\Big|_{(x_0,t')} + 2\left(\frac{v_c^2\ell^2}{\widetilde{c}^2}\right)\frac{\partial^2\widehat{G}}{\partial x^2}\Big|_{(x_0,t')}\right]dn_{c\vartheta}$$

$$(1.34)$$

where v_c denotes the mean speed of the molecules having velocity in the interval I_c.

In order to derive from Eq. (1.34) the flux dq_c of $\widehat{G}$ across ω, we have to divide Eq. (1.34) by dt and then integrate the obtained expression on the interval $[0, \pi/2]$. It is immediately seen that the last three terms in Eq. (1.34) do not contribute to dq_c, due to the form of $dn_{c\vartheta}$ in Eq. (1.33). The first three terms, instead, may be easily integrated and yield

$$dq_c = \left[-\left(\frac{\widetilde{\lambda}_c}{\widetilde{c}}\frac{}{3}\right)\frac{\partial\widehat{G}}{\partial x} + \left(\frac{\ell_c^2}{3}\right)\frac{\partial^2\widehat{G}}{\partial x\partial t} + \left(v_c\frac{\ell_c^2}{3}\right)\frac{\partial^2\widehat{G}}{\partial x^2}\right]dN_c. \qquad (1.35)$$

Let us observe that, being the couple (x_0, t') completely arbitrary, we have omitted such a dependency in deriving Eq. (1.35). By means of the relation $\widehat{G} = 3k_BT/2$, with k_B being the Boltzmann constant, the integration of that relation in the interval $[\widetilde{c} = 0, \widetilde{c} = \infty[$ yields the following total flux of G across ω

$$q = -\lambda\frac{\partial T}{\partial x} + \sigma\frac{\partial^2 T}{\partial x\partial t} + v\sigma\frac{\partial^2 T}{\partial x^2} \qquad (1.36)$$

wherein

$$\lambda = \int_0^{\infty}\frac{\lambda_c c k_B}{2}dN_c$$

$$\sigma = \int_0^{\infty}\frac{l_c^2 k_B}{2}dN_c$$

and c is the mean speed of all the molecules passing across the plane x at the time t, which may be identified with the velocity of the fluid in the point (x, t).

Equation (1.36) can be also written in the form

$$q = -\lambda\frac{\partial}{\partial x}\left(T - \frac{\sigma}{\lambda}\dot{T}\right) \qquad (1.37)$$

where

$$\dot{T} = \frac{\partial T}{\partial t} + v\frac{\partial T}{\partial x}$$

denotes the convective time derivative. Thus, if we define a dynamical temperature β [24, 28] as

$$\beta = T - \frac{\sigma}{\lambda}\dot{T} \tag{1.38}$$

then Eq. (1.37) takes the usual Fourier form

$$q = -\lambda\frac{\partial\beta}{\partial x}. \tag{1.39}$$

The second term in the right-hand side of Eq. (1.38) is an inertial term due to the Van der Waals forces between the molecules, allowing the temperature disturbances to propagate with a finite speed. In fact, the coupling of Eq. (1.38) with its time derivative yields

$$\tau\dot{\beta} + \beta = T - \tau^2\ddot{T} \tag{1.40}$$

where the quantity $\tau = \sigma/\lambda$ may be interpreted as a relaxation time.

In order to evaluate the last term in the right-hand side of Eq. (1.40), we recall that both the gas is in quasi-equilibrium, and in usual applications τ is very small. From the mathematical point of view, these circumstances may be expressed by admitting that if f is a physical quantity related to the heat conduction, then the ratio $|\tau\dot{f}/f|$ is a first-order quantity, and second-order quantities are negligible. In this case we may compare the order of magnitude of $\tau^2\ddot{T}$ with respect to that of T, getting so

$$\left|\tau^2\frac{\ddot{T}}{T}\right| = \left|\tau\frac{\ddot{T}}{\dot{T}}\right|\left|\frac{\tau\dot{T}}{T}\right|$$

which proves that the term $\tau^2\ddot{T}$ in Eq. (1.40) is a second-order quantity. Neglecting it, we conclude that the evolution of β is governed by the linear differential equation

$$\tau\dot{\beta} + \beta = T \tag{1.41}$$

which reveals that for vanishing τ the dynamical temperature β reduces to the absolute one T.

Taking the spatial derivative of Eq. (1.41) and multiplying it by $-\lambda$, one obtains the following evolution equation for the heat flux

$$\tau\dot{q} + q\left(1 + \tau\frac{\partial v}{\partial x}\right) = -\lambda\frac{\partial T}{\partial x}. \tag{1.42}$$

For fluids at rest, as well as for a rigid body, it reduces to the one-dimensional version of the MCV equation (1.6) whenever in Eq. (1.41) $\tau \equiv \tau_R$.

Finally, in the three-dimensional case, Eq. (1.39) may be easily generalized as [24, 28]

$$\mathbf{q} = -\lambda\nabla\beta \tag{1.43}$$

which represents a generalized Fourier-type heat-conduction equation with a dynamical temperature.

1.4.2 Flux-Dependent Absolute Temperature

In EIT the relation between the thermodynamic absolute temperature T, obtained from Eqs. (1.10) and (1.13), and the local-equilibrium one T_{eq}, defined in terms of the internal energy density, without reference to entropy, is given by [17, 30]

$$T = T_{\text{eq}} + \tau\frac{T_{\text{eq}}^2}{2\lambda}\left[\frac{\partial}{\partial u}\left(\frac{1}{T_{\text{eq}}^2}\right)\right]\mathbf{q}\cdot\mathbf{q} \tag{1.44}$$

if the relaxation time τ and the thermal conductivity λ are constant. In microscopic terms, it was argued that T corresponds to the kinetic temperature in the plane perpendicular to the heat flux, and $T \leq T_{\text{eq}}$, whereas T_{eq} corresponds to the total kinetic energy. In other terms, the presence of a heat flux breaks the energy equipartition in the several spatial dimensions, and makes that the kinetic temperature is different along the direction of the heat flux than in the perpendicular directions.

The coupling of Eqs. (1.41) and (1.44) yields

$$\tau\dot{\beta} + \beta = T_{\text{eq}} + \tau\frac{T_{\text{eq}}^2}{2\lambda}\left[\frac{\partial}{\partial u}\left(\frac{1}{T_{\text{eq}}^2}\right)\right]\mathbf{q}\cdot\mathbf{q} \tag{1.45}$$

which further enlightens the dynamical character of β. In this case β and T coincide when β is stationary. It is worth observing that for stationary values of β a heat flux is still present in the system, as it can be inferred by Eq. (1.45). Thus, $\dot{\beta} = 0$ does not imply that the system is in a stationary state. In the absence of memory effects, i.e., if $\tau = 0$, then $\beta = T_{\text{eq}}$.

More complex situations, corresponding to different evolution equations for β are also possible. For instance, in rigid bodies the evolution of β can be ruled by the partial differential equation [30, 31]

$$\tau \dot{\beta} + \beta = T + \frac{\lambda}{2} \left[\frac{\partial \ln (s_\beta)}{\partial u} \right] \nabla \beta \cdot \nabla \beta \qquad (1.46)$$

with

$$s_\beta = \frac{\tau_R}{\lambda T^2}. \qquad (1.47)$$

In this case, the dynamical temperature β and the absolute temperature T do not coincide, even for $\dot{\beta} = 0$.

At the very end, let us observe that the simultaneous presence of u and β in Eqs. (1.45) and (1.46) should be not surprising. Previous considerations suggest that in an equilibrium system, where the volumetric internal energy u is only a function of the local-equilibrium temperature, the use of both u and β would be redundant. Out of equilibrium, instead, since u shows a different distribution, β is a truly independent quantity, and not redundant with u. For instance, the difference between β and T could be related to the difference in the average global kinetic energy and the average kinetic energy in the plane orthogonal to $\mathbf{q}$.

1.4.3 Comparison of Nonequilibrium Temperatures Through Heat Pulse Experiments

A recent theory of heat conduction in isotropic rigid bodies [60], still in the framework of EIT, replaces the classical state-space variables of EIT with two other new variables. Such theory partially departs from the classical assumptions of EIT, since it assumes that the state space is spanned by the dynamical temperature β (which replaces the absolute temperature T) together with a new flux variable $\mathbf{c}$ (which replaces the classical heat flux $\mathbf{q}$).

In this new theory, the former state-space variable is ruled by Eq. (1.41), and the latter state-space variable is ruled by the following evolution equation [60]:

$$\dot{\mathbf{c}} = -\nabla \beta + \frac{2\lambda U_0^2}{\beta} \left[\nabla^T \mathbf{c} + \nabla \mathbf{c} + (\nabla \cdot \mathbf{c}) \mathbf{I} \right] \cdot \mathbf{c} - \frac{\mathbf{c}}{\tau_R} \qquad (1.48)$$

with

$$U_0 = \sqrt{\frac{\lambda}{\tau_R c_v}} \qquad (1.49)$$

being the speed of propagation of thermal pulses traveling through an equilibrium state. The vector $\mathbf{c}$ is a renormalized flux variable which is related to $\mathbf{q}$ through a suitable constitutive equation, and only depends on the dynamical quantity τ_R and on the temperature gradient, but not on other material quantities. Thus, though being very similar to $\mathbf{q}$, it presents qualitative differences with $\mathbf{q}$, and the formalism based on it deserves a comparative analysis with that based on the heat flux, allowing so a deeper understanding of the nonequilibrium subspace of the basic independent variables of the theory.

In order to go deeper into the theory, let us postulate the following constitutive equations for the internal energy per unit volume and for the heat flux in terms of β and $\mathbf{c}$

$$u = u\left(\beta, c^2\right) \tag{1.50a}$$

$$\mathbf{q} = \left(\frac{\lambda}{\tau_R}\right) \mathbf{c}. \tag{1.50b}$$

It is worth observing that Eq. (1.50b) is very similar to the well-known relationship $\mathbf{p} = \mathbf{q}/\bar{v}^2$ appearing in Grad's theory [45], being $\mathbf{p}$ the first moment of the distribution function, and $\bar{v}$ the modulus of the average phonon speed (that is, the first sound).

Then, let us consider the propagation of a smooth surface described by the following equation:

$$\phi\left(\mathbf{x}, t\right) = 0 \tag{1.51}$$

across which the state variables are continuous, but their first-order spatial derivatives suffer jump discontinuities defined by

$$\delta = \left(\frac{\partial}{\partial \phi}\right)_{\phi=0^+} - \left(\frac{\partial}{\partial \phi}\right)_{\phi=0^-}. \tag{1.52}$$

For the sake of simplicity, let us put ourselves in the hypothesis of constant material functions and consider only one-dimensional waves, namely, $\beta \equiv \beta(x, t)$ and $c \equiv c(x, t)$, being x the direction of propagation. Making use of the standard transformations [85]

$$\dot{f} \rightarrow -U\delta f, f_{,x} \rightarrow \delta f \tag{1.53}$$

being U the speed of propagation of the wave, from the local balance of energy per unit volume (1.8) we get

$$\delta\left[u^\beta \dot{\beta} + 2u^\star c\dot{c} + \frac{\lambda}{\tau_R} c_{,x}\right] = 0 \Rightarrow Uu^\beta \delta\beta + \left(2Ucu^\star - \frac{\lambda}{\tau_R}\right)\delta c = 0 \tag{1.54}$$

once the dependence of u on the state-space variables has been explicitly expressed, accordingly with Eq. (1.50a). Note that in the equation above $u^\star = \partial u/\partial c^2$ and $u^\beta = \partial u/\partial \beta$, just for the sake of a compact notation.

From the other hand, the evolution equation (1.48) of $\mathbf{c}$ gives

$$\delta\left[\dot{c} + \left(1 + 3\frac{\lambda c^2}{\beta^2}U_0^2\right)\beta_{,x} - 6\frac{c\lambda}{\beta}U_0^2 c_{,x} + \frac{c}{\tau_R}\right] = 0$$

$$\Rightarrow \left(1 + 3\frac{\lambda c^2}{\beta^2}U_0^2\right)\delta\beta - \left(U + 6\frac{c\lambda}{\beta}U_0^2\right)\delta c = 0. \qquad (1.55)$$

The linear system of Eqs. (1.54) and (1.55) has nontrivial solutions provided the following characteristic equation is satisfied

$$U^2 u^\beta + 2c\left[3u^\beta\frac{\lambda}{\beta}U_0^2 + u^\star\left(1 + 3\frac{\lambda c^2}{\beta^2}U_0^2\right)\right]U - \frac{\lambda}{\tau_R}\left(1 + 3\frac{\lambda c^2}{\beta^2}U_0^2\right) = 0$$

$$(1.56)$$

in virtue of which one may conclude that Eqs. (1.8) and (1.48) constitute a hyperbolic system if, and only if,

$$\Delta \equiv c^2\left[3u^\beta\frac{\lambda}{\beta}U_0^2 + u^\star\left(1 + 3\frac{\lambda c^2}{\beta^2}U_0^2\right)\right]^2 + u^\beta\frac{\lambda}{\tau_R}\left(1 + 3\frac{\lambda c^2}{\beta^2}U_0^2\right) > 0. \quad (1.57)$$

The characteristic velocities arising from Eq. (1.56) are given by [60]

$$U^\pm(\beta, c^2) = \frac{c\left[3u^\beta\frac{\lambda}{\beta}U_0^2 - u^\star\left(1 - 3\frac{\lambda c^2}{\beta^2}U_0^2\right)\right] \pm \sqrt{\Delta}}{u^\beta}, \qquad (1.58)$$

wherein the upper sign $+$ denotes the wave traveling in the same direction of $\mathbf{c}$ (or, equivalently, of the heat flux), and the upper sign $-$ denotes the wave which travels in the opposite direction.

For systems in which the internal energy is independent of the heat flux, the relation above reduces to

$$U^\pm(\beta, c^2) = \frac{3cu^\beta\frac{\lambda}{\beta}U_0^2 \pm \sqrt{\Delta}}{u^\beta}. \qquad (1.59)$$

From Eq. (1.59) it follows that U^+ is higher than U^-, since $U^+ - U^- = 2\sqrt{\Delta}/u^\beta$. Finally, for waves traveling through an equilibrium state, which is characterized by $\mathbf{c} = 0$, Eq. (1.58) simply gives $U_{eq}^\pm = \pm U_0$. It is worth observing that, under the hypothesis $u^\star = 0$, and close to the equilibrium of β, i.e., under the approximation

$u^\beta \simeq c_v = \partial u/\partial T$ and $\beta \simeq T$, we get

$$\Delta \equiv c^2 \left(3c_v \frac{\lambda}{T} U_0^2 \right)^2 + c_v \frac{\lambda}{\tau_R} \left(1 - 3\frac{\lambda c^2}{T^2} U_0^2 \right). \tag{1.60}$$

For dielectric crystals at low temperature, in which the hypothesis $u^\star = 0$ seems to be well-suited, there is a critical temperature T_c at which second-sound propagation takes place. For instance, $T_c \simeq 16.5\,K$ for Sodium Fluoride [51], and $T_c \simeq 4.1\,K$ for Bismuth [73]. On the other hand, for these crystals the material functions λ, c_v and τ_R, as well as the equilibrium wave speed U_0, are well-known functions of temperature, so that the constant value of λ, c_v and τ_R, and the value of U_0 can be taken equal to $\lambda(T_c)$, $c_v(T_c)$, $\tau_R(T_c)$ and $U_0(T_c)$, respectively. Hence, from Eq. (1.60), it follows that the value of Δ and, as a consequence, the quantity $U^+ - U^- = 2\sqrt{\Delta}/c_v$ at the critical temperature, can be obtained by measurements of the heat flux only.

On the other hand, a similar (but not equal) expression of Δ has been obtained by Lebon et al. [63] under the assumption that T, instead of β, enters the state space. These authors obtained $U^+ - U^- = \sqrt{\Delta}/c_v$. Then, a direct measurement of $U^+ - U^-$ could allow to infer whether, out of equilibrium, T or β drives the heat flow in dielectric crystals at low temperature. Although such an experiment is not easy to realize, since it requires very pure specimens at temperatures close to the absolute zero, it is worth to do it, since it is related to an important concept, namely, the definition of temperature in nonequilibrium situations.

1.4.4   Propagation of Temperature Waves Along Cylindrical Nanowires

An alternative way to compare the two temperatures T and β is to study the propagation of heat waves along cylindrical nanowires which are not laterally isolated [60]. Assuming as it usual that the absolute temperature is related with the internal energy per unit volume by means of the relation $du = c_v dT$, then the evolution equation for T in a cylinder of radius r is

$$c_v \dot{T} = -\nabla \cdot \mathbf{q} - \frac{2\sigma_{exc}}{r} (T - T_{env}) \tag{1.61}$$

with the second term in the right-hand side accounting for the heat flux exchanged with the surrounding environment across the lateral walls of the system. It is in accordance with the Newton cooling law, being σ_{exc} a suitable heat exchange coefficient, and T_{env} the temperature of the environment, which will be assumed to be constant and homogeneous. Furthermore, in Eq. (1.61) $\mathbf{q}$ stands for the longitudinal heat flux along the length of the cylinder.

The combination of Eqs. (1.6) and (1.61) turns out

$$\ddot{T} + \left(\frac{1}{\tau_R} + \frac{1}{\tau_r} \right) \dot{T} = U_0^2 \nabla^2 T - \left(\frac{T - T_{\text{env}}}{\tau_R \tau_r} \right) \qquad (1.62)$$

with

$$\tau_r = \frac{r c_v}{2 \sigma_{\text{exc}}}$$

as the relaxation time due to the lateral dispersion of heat, which expresses the time-lag due to the heat exchange of the lateral mantle only. It is not related to collisions' processes between phonons, which are all incorporated in the relaxation time τ_R. It is possible to see from Eq. (1.62) that, due to the small thickness of the conductor, the lateral dispersion introduces a competitive time scale, which influences the evolution of T. In the limits of large radius (i.e., when $r \to \infty$), or for isolated wires (i.e., if $\sigma_{\text{exc}} = 0$), $\tau_r \to \infty$ and Eq. (1.62) reduces to the classical telegraph equation arising in MCV theory.

In Fig. 1.1 we plot at different scales the ratio τ_R/τ_r in a silicon nanowire, as a function of the temperature, for three different values of the radius. From that figure it is possible to see that in the range of temperatures from 100 to 300 K, τ_R is of the order of $10^{-5}\tau_r$: hence in this temperature range, in Eq. (1.62) the term $\dot{T}/\tau_R$ is predominant with respect to $\dot{T}/\tau_r$. In the range from 30 to 100 K, instead, τ_R is of

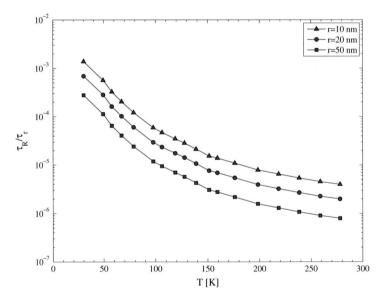

Fig. 1.1 Behavior of the ratio τ_R/τ_r as a function of temperature in a silicon nanowire, for three different values of the radius, r = 10 nm, r = 20 nm, and r = 50 nm. The y axis is in a logarithmic scale in figure

the order of $10^{-3}\tau_r$, so that the influence $\dot{T}/\tau_r$ on the solution of Eq. (1.62) may be relevant.

In computing the relaxation time τ_r, the specific heat per unit volume c_v has been obtained by the Debye expression

$$c_v = \frac{12\pi^4}{5}\left(\frac{T}{T_B}\right)^3\left(\frac{R\rho}{M}\right)$$

with T_B as the Debye temperature (for silicon it holds $T_B = 645$ K), $R = 8.31$ JK^{-1} mol^{-1} as the gas constant, M the molar mass (for silicon $M = 28 \cdot 10^{-3}$ Kg mol^{-1}), and ρ the mass density (for silicon $\rho = 2.33 \cdot 10^3$ Kg m^{-3}). For the heat exchange coefficient σ_{exc} we have taken $\sigma_{\mathrm{exc}} = 4\sigma_{SB}\theta^3$, being $\sigma_{SB} = 5.67 \cdot 10^{-8}$ Wm^{-2} K^{-4} the Stefan-Boltzmann constant (i.e., we have assumed a radiative heat exchange as dominant mechanism). Moreover the resistive relaxation time has been estimated as $\tau_R = \ell_p/\bar{v}$.

For silicon and for germanium, at the different temperatures, the values of the bulk thermal conductivity and of the phonon mean-free path are summarized in Tables 1.1 and 1.2. Note that the latter values have been obtained from the relation $\lambda = (1/3)c_v\bar{v}\ell_p$. In fact, the phonon mean-free path depends both on the phonon frequency, and on the kind of collisions, in such a way several different relevant averages may be used for it. Along this book, we will often use that definition for ℓ_p.

Consider now the propagation of the following plane temperature-wave from a stationary reference level $\bar{T}(x)$

$$T(x;t) = \bar{T}(x) + \bar{T}_0 e^{[i(\omega t - \kappa x)]} \tag{1.63}$$

where x means the longitudinal spatial coordinate, ω is the frequency of perturbation, and κ is the wave number. From the experimental point of view, a similar signal may be realized by imposing at one end of the system a sinusoidally time-dependent temperature, and detecting the consequent temperature perturbation at different points along the system.

When Eq. (1.63) is inserted into Eq. (1.62), the following dispersion relation is derived

$$\kappa^2 = \frac{1}{d^2}\left(\omega^2\tau_R^2 - \frac{\tau_R}{\tau_r}\right) - i\frac{\omega\tau_R}{d^2}\left(1 + \frac{\tau_R}{\tau_r}\right) \tag{1.64}$$

with $d \equiv U_0\tau_R$ as a suitable length. The corresponding phase velocity $U_p \equiv |\omega/\operatorname{Re}(\kappa)|$ and the attenuation distance $\alpha \equiv |1/\operatorname{Im}(\kappa)|$ are, respectively,

$$U_{p/T} = \frac{\omega}{\Gamma_T}\sqrt{\frac{2}{1+\phi_T}} \tag{1.65a}$$

$$\alpha_{/T} = \frac{1}{\Gamma_T}\sqrt{\frac{2}{1-\phi_T}} \tag{1.65b}$$

Table 1.1 Bulk thermal conductivity λ $\left(\mathrm{Wm^{-1}\,K^{-1}}\right)$, specific heat c_v $\left(\mathrm{Jm^{-3}\,K^{-1}}\right)$, average phonon mean-free path ℓ_p (m) and average phonon speed $\bar{v}$ $\left(\mathrm{ms^{-1}}\right)$ obtained from experimental data, for silicon at different temperatures T (K)

T	λ	c_v	ℓ_p	$\bar{v}$
30	3097.5	53,068.65	$4.93 \cdot 10^{-5}$	3554.38
35	3046.8	85,975.47	$3.03 \cdot 10^{-5}$	3499.07
40	2866.49	124,456.63	$1.98 \cdot 10^{-5}$	3482.16
45	2616.42	166,249.35	$1.35 \cdot 10^{-5}$	3483.05
50	2342.46	209,605.33	$9.60 \cdot 10^{-6}$	3491.54
55	2074.82	253,383.03	$7.01 \cdot 10^{-6}$	3502.26
60	1829.81	296,952.43	$5.26 \cdot 10^{-6}$	3512.11
65	1613.52	340,045.31	$4.04 \cdot 10^{-6}$	3519.22
70	1425.97	382,614.17	$3.17 \cdot 10^{-6}$	3522.52
75	1264.37	424,722.65	$2.54 \cdot 10^{-6}$	3521.48
80	1125.15	466,470.79	$2.06 \cdot 10^{-6}$	3516.00
85	1004.98	507,950.15	$1.69 \cdot 10^{-6}$	3506.30
90	901.02	549,221.13	$1.41 \cdot 10^{-6}$	3492.81
95	810.97	590,305.46	$1.19 \cdot 10^{-6}$	3476.10
100	732.88	631,187.79	$1.01 \cdot 10^{-6}$	3456.78
150	336.27	1,013,219.58	$3.08 \cdot 10^{-7}$	3226.05
200	213.12	1,307,155.56	$1.60 \cdot 10^{-7}$	3057.28
250	158.75	1,510,418.47	$1.07 \cdot 10^{-7}$	2956.42
300	128.03	1,648,423.99	$8.05 \cdot 10^{-8}$	2894.96
350	107.95	1,743,630.51	$6.50 \cdot 10^{-8}$	2855.57
400	93.63	1,811,033.32	$5.48 \cdot 10^{-8}$	2829.05
450	82.82	1,860,059.38	$4.76 \cdot 10^{-8}$	2810.44
500	74.34	1,896,630.49	$4.20 \cdot 10^{-8}$	2796.91
550	67.50	1,924,535.97	$3.77 \cdot 10^{-8}$	2786.78
600	61.84	1,946,260.34	$3.43 \cdot 10^{-8}$	2779.01

The average mean-free path has been determined from the relation $\lambda = (1/3)\,c_v\bar{v}\ell_p$

where

$$
\begin{cases}
\Gamma_T = \dfrac{1}{U_0}\sqrt[4]{\left(\omega^2 - \dfrac{1}{\tau_R\tau_r}\right)^2 + \dfrac{\omega^2}{\tau_R^2}\left(1 + \dfrac{\tau_R}{\tau_r}\right)^2} \\[2ex]
\phi_T = \dfrac{\omega^2\tau_R\tau_r - 1}{\sqrt{\left(1 + \omega^2\tau_R^2\right)\left(1 + \omega^2\tau_r^2\right)}}
\end{cases}
\tag{1.66}
$$

and the subscript (T) means that in the Newton cooling law we considered the contribution of the absolute temperature T.

At low frequencies (lf), i.e., when $\tau_R\omega \ll 1$, it is found that

$$
U_{p/T}^{(\mathrm{lf})} = 2U_0\left[1 - \frac{\omega^2}{4}\left(\tau_R^2 + \tau_r^2\right)\right]\frac{\sqrt{\tau_R\tau_r}}{\tau_R + \tau_r}
\tag{1.67a}
$$

Table 1.2 Bulk thermal conductivity λ $\left(\text{Wm}^{-1}\,\text{K}^{-1}\right)$, specific heat c_v $\left(\text{Jm}^{-3}\,\text{K}^{-1}\right)$, average phonon mean-free path ℓ_p (m) and average phonon speed $\bar{v}$ $\left(\text{ms}^{-1}\right)$ obtained from experimental data, for germanium at different temperatures T (K)

T	λ	c_v	ℓ_p	$\bar{v}$
30	1039.15	212,101.62	$7.04 \cdot 10^{-6}$	2088.38
35	944.83	279,811.52	$4.83 \cdot 10^{-6}$	2096.44
40	832.10	345,681.31	$3.43 \cdot 10^{-6}$	2102.05
45	718.14	409,892.56	$2.50 \cdot 10^{-6}$	2102.64
50	613.61	473,008.68	$1.85 \cdot 10^{-6}$	2097.48
55	523.27	535,405.25	$1.41 \cdot 10^{-6}$	2087.10
60	447.92	597,139.60	$1.08 \cdot 10^{-6}$	2072.69
65	386.30	658,013.64	$8.57 \cdot 10^{-7}$	2055.57
70	336.37	717,687.48	$6.90 \cdot 10^{-7}$	2036.98
75	296.02	775,780.08	$5.67 \cdot 10^{-7}$	2017.89
80	263.33	831,937.50	$4.75 \cdot 10^{-7}$	1998.99
85	236.72	885,870.26	$4.05 \cdot 10^{-7}$	1980.76
90	214.89	937,367.58	$3.50 \cdot 10^{-7}$	1963.47
95	196.84	986,296.92	$3.07 \cdot 10^{-7}$	1947.28
100	181.77	1,032,595.58	$2.73 \cdot 10^{-7}$	1932.24
150	108.84	1,364,731.4	$1.30 \cdot 10^{-7}$	1834.68
200	82.26	1,536,601.36	$8.96 \cdot 10^{-8}$	1791.58
250	67.48	1,630,686.57	$7.01 \cdot 10^{-8}$	1769.93
300	57.65	1,686,389.91	$5.83 \cdot 10^{-8}$	1757.70
350	50.50	1,721,692.39	$5.03 \cdot 10^{-8}$	1750.18
400	45.02	1,745,334.45	$4.43 \cdot 10^{-8}$	1745.22
450	40.67	1,761,886.98	$3.98 \cdot 10^{-8}$	1741.80
500	37.12	1,773,902.52	$3.61 \cdot 10^{-8}$	1739.34
550	34.17	1,782,888.55	$3.31 \cdot 10^{-8}$	1737.51
600	31.66	1,789,778.46	$3.06 \cdot 10^{-8}$	1736.11

The average mean-free path has been determined from the relation $\lambda = (1/3)\,c_v\bar{v}\ell_p$

$$\alpha_{/T}^{(\text{lf})} = d \left[1 - \frac{\omega^2}{8} (\tau_r - \tau_R)^2 \right] \sqrt{\frac{\tau_r}{\tau_R}} \qquad (1.67b)$$

which are the results predicted by Fourier law for an isolated wire, plus contributions in τ_R and τ_r, which vanish in the previous case.

In the high-frequency (hf) limit (that is, when $\tau_R\omega \to \infty$), instead, the phase velocity and the attenuation distance become, respectively,

$$U_{p/T}^{(\text{hf})} = U_0 \qquad (1.68a)$$

$$\alpha_{/T}^{(\text{hf})} = 2d \left(\frac{\tau_r}{\tau_r - \tau_R} \right). \qquad (1.68b)$$

The result (1.68a) for the phase speed is the same as in the bulk and it does not depend on the radius. In contrast, the radius influences the attenuation length (1.68b), and in the limit of high radius, or of isolated nanowires, one recovers the usual result for thermal waves in hyperbolic heat propagation.

Up to here, it has been assumed that the heat transport is related to the gradient of the absolute nonequilibrium temperature T. Indeed, according with Eq. (1.43), the heat transport may be also related to the dynamical temperature β, as well as the term accounting for the thermal exchange with the environment may be related to the difference $(\beta - \beta_{\text{env}})$. In this case Eq. (1.61) has to be replaced by

$$c_v \dot{T} = -\nabla \cdot \mathbf{q} - \frac{2\sigma_{\text{exc}}}{r} (\beta - \beta_{\text{env}}). \tag{1.69}$$

Thus, we have assumed that the thermodynamic quantity leading the heat flux is β, rather than T. This hypothesis is consistent with Eq. (1.43) in the same way that Eq. (1.61) is compatible with Eq. (1.6). Note that we have not substituted T with β into the whole set of constitutive equations, since we still consider u as depending on T. However, since we are assuming that the quantity driving the heat flow is β rather than T, we have substituted T with β into the Newton cooling law.

The comparison of results from Eqs. (1.61) and (1.69) could thus allow to search whether $(T - T_{\text{env}})$, or $(\beta - \beta_{\text{env}})$ provide the best description of heat transfer in fast processes.

Assuming that in the environment the relaxation effects are absent (i.e., $\tau \dot{\beta} = 0$) by Eq. (1.41) it follows that $T_{\text{env}} \equiv \beta_{\text{env}}$, which is assumed to be constant, as well as in Eq. (1.61). Combining Eqs. (1.43) and (1.69) one obtains now

$$\ddot{T} + \frac{\dot{T}}{\tau_R} = U_0^2 \nabla^2 T - \left(\frac{T - T_{\text{env}}}{\tau_R \tau_r} \right) \tag{1.70}$$

instead of Eq. (1.62). It is important to note the difference between Eqs. (1.62) and (1.70), since the term in $\dot{T}/\tau_R$ in the round bracket of the left-hand side of Eq. (1.62) is absent in Eq. (1.70).

Assuming again Eq. (1.63) for the heat waves, Eq. (1.70) leads to the following new dispersion relation

$$\kappa^2 = \frac{1}{d^2} \left(\omega^2 \tau_R^2 - \frac{\tau_R}{\tau_r} \right) - i \frac{\omega \tau_R}{d^2} \tag{1.71}$$

yielding for the phase velocity and the attenuation distance, respectively,

$$U_{p/\beta} = \frac{\omega}{\Gamma_\beta} \sqrt{\frac{2}{1 + \phi_\beta}}, \tag{1.72a}$$

$$\alpha_{/\beta} = \frac{1}{\Gamma_\beta} \sqrt{\frac{2}{1 - \phi_\beta}} \tag{1.72b}$$

with

$$\begin{cases} \Gamma_\beta = \dfrac{1}{U_0} \sqrt[4]{\left(\omega^2 - \dfrac{1}{\tau_R \tau_r}\right)^2 + \dfrac{\omega^2}{\tau_R^2}} \\[2ex] \phi_\beta = \dfrac{\omega^2 \tau_R \tau_r - 1}{\sqrt{\left(\omega^2 \tau_R \tau_r - 1\right)^2 + \omega^2 \tau_r^2}} \end{cases} \qquad (1.73)$$

and the subscript (β) indicating that now, in the Newton cooling law, we considered the contribution of the dynamical temperature β, in contrast with previous analysis, wherein T was used.

At low frequency the phase velocity and the attenuation distance are given by

$$U_{p/\beta}^{(lf)} = 2U_0 \left[1 - \frac{\omega^2}{2}\left(\tau_r^2 - 2\tau_R \tau_r\right)\right]\sqrt{\frac{\tau_R}{\tau_r}} \qquad (1.74a)$$

$$\alpha_{/\beta}^{(lf)} = 0d\left[1 - \frac{\omega^2}{8}\left(\tau_r^2 - \tau_R \tau_r\right)\right]\sqrt{\frac{\tau_r}{\tau_R}}. \qquad (1.74b)$$

In the high-frequency limit one has, instead,

$$U_{p/\beta}^{(hf)} = U_0 \qquad (1.75a)$$

$$\alpha_{/\beta}^{(hf)} = 2d \qquad (1.75b)$$

which coincide with the results of heat waves in the bulk or in isolated systems. Note that Eq. (1.75b) does not depend on τ_r, in contrast, with Eq. (1.72b).

The relative difference of the phase velocities Eqs. (1.68a) and (1.74a) for very small frequency is

$$\frac{U_{p/\beta}^{(lf)} - U_{p/T}^{(lf)}}{U_{p/\beta}^{(lf)}} = \frac{\dfrac{\tau_R}{\tau_r}}{1 + \dfrac{\tau_R}{\tau_r}}. \qquad (1.76)$$

Thus, the essential parameter reflecting this difference is the nondimensional ratio τ_R/τ_r, which for low temperature is $\tau_R/\tau_r \approx 2.92 \cdot 10^{-9}$ Kn. At room temperature, the Knudsen number gets a very small value, but at low temperature, it increases since the mean-free path ℓ_p becomes very large (see Table 1.1). That way the ratio τ_R/τ_r gets the order of 10^{-5} at temperatures of the order of 30 K. Since the phase speed, in general, is of the order of $10^3 \, \text{ms}^{-1}$, the difference $\left(U_{p/\beta} - U_{p/T}\right)$ may be of the order of some cm s^{-1}. This difference could be measured, in principle. Furthermore, in the presence of air, adding the convective effects to the purely radiative effects in the heat exchange with the environment, may give a more important contribution [60].

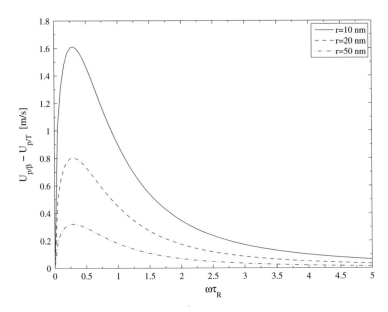

Fig. 1.2 Behavior of the difference between $U_{p/\beta}$ and $U_{p/T}$ in a silicon nanowire at 30 K as a function of $\omega\tau$, for three different radii (i.e., r = 10 nm, r = 20 nm and r = 50 nm)

Figure 1.2 shows, for three different radii, the difference between the two phase velocities $U_{p/\beta}$ and $U_{p/T}$ in a nanowire as a function of $\omega\tau_R$. The nanowire is made of silicon and is at 30 K. In all cases, the difference gets its maximum value for $\omega\tau_R \approx 0.3$, and may be of the order of 1.5 ms^{-1} for r = 10 nm.

References

1. Alvarez, F.X., Jou, D.: Memory and nonlocal effects in heat transports: from diffusive to ballistic regime. Appl. Phys. Lett. **90**, 083109 (3 pp.) (2007)
2. Alvarez, F.X., Jou, D.: Size and frequency dependence of effective thermal conductivity in nanosystems. J. Appl. Phys. **103**, 094321 (8 pp.) (2008)
3. Balandin, A.A.: Thermal properties of graphene and nanostructured carbon materials. Nat. Mater. **10**, 569–581 (2011)
4. Balandin, A., Wang, K.L.: Significant decrease of the lattice thermal conductivity due to phonon confinement in a free-standing semiconductor quantum well. Phys. Rev. B **58**, 1544–1549 (1998)
5. Balandin, A.A., Ghosh, S., Baoand, W., Calizo, I., Teweldebrhan, D., Miao, F., Lau, C.-N.: Superior thermal conductivity of single-layer graphene. Nano Lett. **8**, 902–907 (2008)
6. Banach, Z., Larecki, W.: Nine-moment phonon hydrodynamics based on the modified Grad-type approach: formulation. J. Phys. A: Math. Gen. **37**, 9805–9829 (2004)
7. Banach, Z., Larecki, W.: Nine-moment phonon hydrodynamics based on the maximum-entropy closure: one-dimensional flow. J. Phys. A: Math. Gen. **38**, 8781–8802 (2005)

8. Banach, Z., Larecki, W.: Chapman-Enskog method for a phonon gas with finite heat flux. J. Phys. A: Math. Gen. **41**, 375502 (18 pp.) (2008)
9. Benedetto, G., Boarino, L., Spagnolo, R.: Evaluation of thermal conductivity of porous silicon layers by a photoacoustic method. Appl. Phys. A: Mater. Sci. Process. **64**, 155–159 (1997)
10. Benedict, L.X., Louie, S.G., Cohen, M.L.: Heat capacity of carbon nanotubes. Solid State Commun. **100**, 177–180 (1996)
11. Boltzmann, L.: Leçons sur la Théorie des Gaz. Gauthier-Villars, Paris (1902)
12. Boukai, A.I., Bunimovich, Y., Tahir-Kheli, J., Yu, J.-K., Goddard-III, W.A., Heath, J.R.: Silicon nanowires as efficient thermoelectric materials. Nature **451**, 168–171 (2008)
13. Cahill, D.C., Ford, W.K., Goodson, K.E., Mahan, G.D., Majumdar, A., Maris, H.J., Merlin, R., Phillpot, S.R.: Nanoscale thermal transport. J. Appl. Phys. **93**, 793–818 (2003)
14. Cahill, D.G., et al.: Nanoscale thermal transport. II. 2003–2012. Appl. Phys. Rev. **1**, 011305 (45 pp.) (2014)
15. Canham, L.T.: Silicon quantum wire fabrication by electrochemical and chemical dissolution of wafers. Appl. Phys. Lett. **57**, 1046–1048 (1990)
16. Cao, B.-Y., Guo, Z.-Y.: Equation of motion of a phonon gas and non-Fourier heat conduction. J. Appl. Phys. **102**, 053503 (6 pp.) (2007)
17. Casas-Vázquez, J., Jou, D.: Temperature in nonequilibrium states: a review of open problems and current proposals. Rep. Prog. Phys. **66**, 1937–2023 (2003)
18. Cattaneo, C.: Sulla conduzione del calore. Atti Sem. Mat. Fis. Univ. Modena **3**, 83–101 (1948)
19. Cattaneo, C.: Sur une forme de l'équation de la chaleur éliminant le paradoxe d'une propagation instantanée. C. R. Acad. Sc. **247**, 431–433 (1958)
20. Chen, G.: Ballistic-diffusion equations for transient heat conduction from nano to macroscales. J. Heat Transf. - T. ASME **124**, 320–328 (2001)
21. Chen, G.: Ballistic-diffusive heat-conduction equations. Phys. Rev. Lett. **86**, 2297–2300 (2001)
22. Chen, G.: Nanoscale Energy Transport and Conversion - A Parallel Treatment of Electrons, Molecules, Phonons, and Photons. Oxford University Press, Oxford (2005)
23. Chung, J.D.,, Kaviany, M.: Effects of phonon pore scattering and pore randomness on effective conductivity of porous silicon. Int. J. Heat Mass Transf. **43**, 521–538 (2000)
24. Cimmelli, V.A.: Mesoscopic approach to inviscid gas dynamics with thermal lag. Ann. Phys. (Berlin) **525**, 921–933 (2013)
25. Cimmelli, V.A., Frischmuth, K.: Determination of material functions through second sound measurements in a hyperbolic heat conduction theory. Math. Comput. Model. **24**, 19–28 (1996)
26. Cimmelli, V.A., Frischmuth, K.: Nonlinear effects in thermal wave propagation near zero absolute temperature. Physica B **355**, 147–157 (2005)
27. Cimmelli, V.A., Frischmuth, K.: Gradient generalization to the extended thermodynamic approach and diffusive-hyperbolic heat conduction. Physica B **400**, 257–265 (2007)
28. Cimmelli, V.A., Kosiński, W.: Non-equilibrium semi-empirical temperature in materials with thermal relaxation. Arch. Mech. **47**, 753–767 (1991)
29. Cimmelli, V.A., Ván, P.: The effects of nonlocality on the evolution of higher order fluxes in nonequilibrium thermodynamics. J. Math. Phys. **46**, 112901 (15 pp.) (2005)
30. Cimmelli, V.A., Sellitto, A., Jou, D.: Nonlocal effects and second sound in a nonequilibrium steady state. Phys. Rev. B **79**, 014303 (13 pp.) (2009)
31. Cimmelli, V.A., Sellitto, A., Jou, D.: Nonequilibrium temperatures, heat waves, and nonlinear heat transport equations. Phys. Rev. B **81**, 054301 (9 pp.) (2010)
32. Cimmelli, V.A., Sellitto, A., Jou, D.: Nonlinear evolution and stability of the heat flow in nanosystems: beyond linear phonon hydrodynamics. Phys. Rev. B **82**, 184302 (9 pp.) (2010)
33. Cimmelli, V.A., Jou, D., Ruggeri, T., Ván, P.: Entropy principle and recent results in non-equilibrium theories. Entropy **16**, 1756–1807 (2014)
34. Coleman, B.D., Mizel, V.J.: On the general theory of fading memory. Arch. Ration. Mech. Anal. **29**, 18–31 (1968)
35. Criado-Sancho, J.M., Jou, D., Casas-Vázquez, J.: Nonequilibrium kinetic temperatures in flowing gases. Phys. Lett. A **350**, 339–341 (2006)

36. de Groot, S.R., Mazur, P.: Nonequilibrium Thermodynamics. North-Holland Publishing Company, Amsterdam (1962)
37. Dedeurwaerdere, T., Casas-Vázquez, J., Jou, D., Lebon, G.: Foundations and applications of a mesoscopic thermodynamic theory of fast phenomena. Phys. Rev. E **53**, 498–506 (1996)
38. Demirel, Y., Sandler, I.: Linear-nonequilibrium thermodynamics theory for coupled heat and mass transport. Int. J. Heat Mass Transf. **44**, 2439–2451 (2001)
39. Dong, Y.: Dynamical Analysis of Non-Fourier Heat Conduction and Its Application in Nanosystems. Springer, Berlin/Heidelberg/New York (2016)
40. Dong, Y., Cao, B.-Y., Guo, Z.-Y.: Generalized heat conduction laws based on thermomass theory and phonon hydrodynamics. J. Appl. Phys. **110**, 063504 (6 pp.) (2011)
41. Dong, Y., Cao, B.-Y., Guo, Z.-Y.: General expression for entropy production in transport processes based on the thermomass model. Phys. Rev. E **85**, 061107 (8 pp.) (2012)
42. Dong, Y., Cao, B.-Y., Guo, Z.-Y.: Temperature in nonequilibrium states and non-Fourier heat conduction. Phys. Rev. E **87**, 032150 (8 pp.) (2013)
43. Dreyer, W., Struchtrup, H.: Heat pulse experiments revisited. Contin. Mech. Thermodyn. **5**, 3–50 (1993)
44. Ferrer, M., Jou, D.: Higher-order fluxes and the speed of thermal waves. Int. J. Heat Mass Transf. **34**, 3055–3060 (1991)
45. Grad, H.: On the kinetic theory of rarefied gases. Commun. Pure Appl. Math. **2**, 331–407 (1949)
46. Grmela, M., Lebon, G., Dauby, P.C., Bousmina, M.: Ballistic-diffusive heat conduction at nanoscale: GENERIC approach. Phys. Lett. A **339**, 237–245 (2005)
47. Guyer, R.A., Krumhansl, J.A.: Solution of the linearized phonon Boltzmann equation. Phys. Rev. **148**, 766–778 (1966)
48. Guyer, R.A., Krumhansl, J.A.: Thermal conductivity, second sound and phonon hydrodynamic phenomena in nonmetallic crystals. Phys. Rev. **148**, 778–788 (1966)
49. Gyarmati, I.: On the wave approach of thermodynamics and some problems of non-linear theories. J. Non-Equilib. Thermodyn. **2**, 236–260 (1977)
50. Győry, E., Márkus, F.: Size dependent thermal conductivity in nano-systems. Thin Solid Films **565**, 89–93 (2014)
51. Jackson, H.E., Walker, C.T.: Thermal conductivity, second sound, and phonon-phonon interactions in NaF. Phys. Rev. Lett. **3**, 1428–1439 (1971)
52. Jou, D., Criado-Sancho, M.: Thermodynamic stability and temperature overshooting in dual-phase-lag heat transfer. Phys. Lett. A **48**, 172–178 (1998)
53. Jou, D., Restuccia, L.: Mesoscopic transport equations and contemporary thermodynamics: an introduction. Contemp. Phys. **52**, 465–474 (2011)
54. Jou, D., Casas-Vázquez, J., Lebon, G.: Extended irreversible thermodynamics revisited (1988–1998). Rep. Prog. Phys. **62**, 1035–1142 (1999)
55. Jou, D., Casas-Vázquez, J., Lebon, G., Grmela, M.: A phenomenological scaling approach for heat transport in nano-systems. Appl. Math. Lett. **18**, 963–967 (2005)
56. Jou, D., Casas-Vázquez, J., Lebon, G.: Extended irreversible thermodynamics of heat transport. A brief introduction. Proc. Est. Acad. Sci. **57**, 118–126 (2008)
57. Jou, D., Cimmelli, V.A., Sellitto, A.: Nonequilibrium temperatures and second-sound propagation along nanowires and thin layers. Phys. Lett. A **373**, 4386–4392 (2009)
58. Jou, D., Casas-Vázquez, J., Lebon, G.: Extended Irreversible Thermodynamics, 4th revised edn. Springer, Berlin (2010)
59. Jou, D., Criado-Sancho, M., Casas-Vázquez, J.: Heat fluctuations and phonon hydrodynamics in nanowires. J. Appl. Phys. **107**, 084302 (4 pp.) (2010)
60. Jou, D., Cimmelli, V.A., Sellitto, A.: Dynamical temperature and renormalized flux variable in extended thermodynamics of rigid heat conductors. J. Non-Equilib. Thermodyn. **36**, 373–392 (2011)
61. Jou, D., Sellitto, A., Alvarez, F.X.: Heat waves and phonon-wall collisions in nanowires. Proc. R. Soc. A **467**, 2520–2533 (2011)

62. Kovács, R., Ván, P.: Generalized heat conduction in heat pulse experiments. Int. J. Heat Mass Transf. **83**, 613–620 (2015)
63. Lebon, G., Ruggieri, M., Valenti, A.: Extended thermodynamics revisited: Renormalized flux variables and second sound in rigid solids. J. Phys.: Condens. Matter **20**, 025223 (11 pp.) (2008)
64. Lebon, G., Jou, D., Casas-Vázquez, J.: Understanding Nonequilibrium Thermodynamics. Springer, Berlin (2008)
65. Lebon, G., Machrafi, H., Grmela, M., Dubois, C.: An extended thermodynamic model of transient heat conduction at sub-continuum scales. Proc. R. Soc. A **467**, 3241–3256 (2011)
66. Lee, S., Broido, D., Esfarjani, K., Chen, G.: Hydrodynamic phonon transport in suspended graphene. Nat. Commun. **6**, 6290 (9 pp.) (2015)
67. Lundstrom, M.: Fundamentals of Carrier Transport. Cambridge University Press, Cambridge (2000)
68. Luzzi, R., Vasconcellos, A.R., Casas-Vázquez, J., Jou, D.: Characterization and measurement of a nonequilibrium temperature-like variable in irreversible thermodynamics. Physica A **234**, 699–714 (1997)
69. Machrafi, H., Lebon, G.: Size and porosity effects on thermal conductivity of nanoporous material with an extension to nanoporous particles embedded in ahost matrix. Phys. Lett. A **379**, 968–973 (2015)
70. Márkus, F., Gambár, K.: Heat propagation dynamics in thin silicon layers. Int. J. Heat Mass Transf. **56**, 495–500 (2013)
71. Minnich, A.J.: Advances in the measurement and computation of thermal phonon transport properties. J. Phys.: Condens. Matter **27**, 053202 (21 pp.) (2015)
72. Müller, I., Ruggeri, T.: Rational Extended Thermodynamics, 2nd edn. Springer, New York (1998)
73. Narayanamurti, V., Dynes, R.D.: Observation of second sound in bismuth. Phys. Rev. Lett. **28**, 1461–1465 (1972)
74. Öttinger, H.C.: Beyond Equilibrium Thermodynamics. Wiley, New York (2005)
75. Pattamatta, A., Madnia, C.K.: Modeling heat transfer in Bi_2Te_3-Sb_2Te_3 nanostructures. Int. J. Heat Mass Transf. **52**, 860–869 (2009)
76. Ruggeri, T., Sugiyama, M.: Rational Extended Thermodynamics Beyond the Monatomic Gas. Springer International Publishing, Switzerland (2015)
77. Sellitto, A., Cimmelli, V.A.: A continuum approach to thermomass theory. J. Heat Transf. - T. ASME **134**, 112402 (6 pp.) (2012)
78. Sellitto, A., Cimmelli, V.A.: Flux limiters in radial heat transport in silicon nanolayers. J. Heat Transf. - T. ASME **136**, 071301 (6 pp.) (2014)
79. Sellitto, A., Alvarez, F.X., Jou, D.: Phonon-wall interactions and frequency-dependent thermal conductivity in nanowires. J. Appl. Phys. **109**, 064317 (8 pp.) (2011)
80. Serdyukov, S.I.: A new version of extended irreversible thermodynamics and dual-phase-lag model in heat transfer. Phys. Lett. A **281**, 16–20 (2001)
81. Serdyukov, S.I.: Higher order heat and mass transfer equations and their justification in extended irreversible thermodynamics. Theor. Found. Chem. Eng. **47**, 89–103 (2013)
82. Serdyukov, S.I., Voskresenskii, N.M., Bel'nov, V.K., Karpov, I.I.: Extended irreversible thermodynamics and generalization of the dual-phase-lag model in heat transfer. J. Non-Equilib. Thermodyn. **28**, 207–219 (2003)
83. Sieniutycz, S.: Conservation Laws in Variational Thermo-Hydrodynamics. Kluwer Academic Publishers, Dordrecht (1994)
84. Smith, G.F.: On isotropic functions of symmetric tensors, skew-symmetric tensors and vectors. Int. J. Eng. Sci. **9**, 899–916 (1971)
85. Straughan, B.: Heat Waves. Springer, Berlin (2011)
86. Struchtrup, H.: Macroscopic Transport Equations for Rarefied Gas Flows: Approximation Methods in Kinetic Theory - Interaction of Mechanics and Mathematics. Springer, New York (2005)

87. Tzou, D.Y.: Nonlocal behavior in phonon transport. Int. J. Heat Mass Transf. **54**, 475–481 (2011)
88. Tzou, D.Y.: Macro- to Microscale Heat Transfer: The Lagging Behaviour, 2nd edn. Wiley, Chichester (2014)
89. Tzou, D.Y., Guo, Z.-Y.: Nonlocal behavior in thermal lagging. Int. J. Therm. Sci. **49**, 1133–1137 (2010)
90. Vasconcellos, A.R., Luzzi, R., Jou, D., Casas-Vázquez, J.: Thermal waves in an extended hydrodynamic approach. Physica A **212**, 369–381 (1994)
91. Vázquez, F., del Río, J.A.: Thermodynamic characterization of the diffusive transport to wave propagation transition in heat conducting thin films. J. Appl. Phys. **112**, 123707 (8 pp.) (2012)
92. Vázquez, F., Márkus, F., Gambár, K.: Quantized heat transport in small systems: A phenomenological approach. Phys. Rev. E **79**, 031113 (7 pp.) (2009)
93. Verhás, J.: On the entropy current. J. Non-Equilib. Thermodyn. **8**, 201–206 (1983)
94. Vernotte, P.: Les paradoxes de la théorie continue de l'équation de la chaleur. C. R. Acad. Sc. **246**, 3154–3155 (1958)
95. Xu, M.T., Wang, L.Q.: Dual-phase lagging heat conduction based on Boltzmann transport equation. Int. J. Heat Mass Transf. **48**, 5616–5624 (2005)
96. Yang, N., Xu, X., Zhang, G., Li, B.: Thermal transport in nanostructures. AIP Adv. **2**, 041410 (24 pp.) (2012)
97. Zhang, Z.M.: Nano/Microscale Heat Transfer. McGraw-Hill, New York (2007)

Chapter 2
Linear and Nonlinear Heat-Transport Equations

Nanotechnology, like biotechnology and information technology, nowadays is a growing industry with the potential to greatly change the world. Engineering of nanosystems rapidly developed in recent years, and it actually allows to design and develop mechanical, optical and electronic devices, the characteristic sizes of which may be of the order of tens of nanometers. Nowadays, nanotechnology is also fighting its way in medicine, offering some exciting possibilities which few years ago were only imagined. Devices operating on nanometer length scale always provide new challenges, especially regarding their thermo-mechanical properties, and researchers face great challenges in thermal management and analysis under the extreme conditions.

The heat-conduction theory, established by Fourier in 1822, prescribes that the local heat flux $\mathbf{q}$ is linearly proportional to the temperature gradient ∇T, and it can only flow from the warmer regions to colder ones [35]. For isotropic materials, such a statement is summarized by Eq. (1.2), i.e., $\mathbf{q} = -\lambda \nabla T$. In literature it is possible to find several theoretical derivations of Eq. (1.2). Based on the kinetic theory of gases, the Boltzmann's proof of the Fourier law (FL) is worth of mention for its clarity and expressivity [7].

Despite its empirical standing in ordinary circumstances, both theoretical and experimental evidences show that Eq. (1.2) no longer holds in several nonequilibrium situations as, for example, heat conduction at nanoscale [16, 33, 40, 43, 51, 57, 75, 81, 86]. The consequent inapplicability of the classical FL in practical situations to well-describe heat transport at nanoscale has led to several generalizations of it [15, 37, 49, 52, 73, 76, 78].

The main features beyond FL are heat waves, ballistic transport, and the so-called phonon hydrodynamics [3–5, 22–24, 67, 69]. Furthermore, as a particular consequence of some of these phenomena, the effective thermal conductivity is no longer only a function of the material, but also of the size and shape of the system.

© Springer International Publishing Switzerland 2016
A. Sellitto et al., *Mesoscopic Theories of Heat Transport in Nanosystems*,
SEMA SIMAI Springer Series 6, DOI 10.1007/978-3-319-27206-1_2

The above features have been much discussed from microscopic bases. Heat waves have also been discussed from macroscopic perspectives, by suitable relaxational extensions of FL, or introducing internal variables, or memory kernels. However, the diffusive-to-ballistic transition, as well as the phonon-hydrodynamic regime have been much studied from a mesoscopic perspective. The formalism we use in this chapter allows to deal with these subjects. Of course, their deep physical interpretation requires to go to microscopic grounds, but it is interesting to have purely mesoscopic generalizations able to make a bridge with the microscopic progress. Since the experimental measurements are usually mesoscopic or macroscopic—as they use the temperature concept, for instance—such a bridge between microscopic and mesoscopic approaches cannot be skipped on practical grounds.

2.1 The Maxwell-Cattaneo-Vernotte Equation and Heat Waves

In the analysis of heat conduction in a non-deformable body at rest, the combination of Eq. (1.2) with the energy-balance equation in absence of source terms (1.8) leads to

$$c_v \dot{T} = \nabla \cdot (\lambda \nabla T) . \tag{2.1}$$

From the mathematical point of view, Eq. (2.1) is a classical example of parabolic partial differential equation. It implies that any thermal disturbance, paradoxically, propagates with infinite speed. In other words, Eq. (2.1) states that the application of a temperature difference gives instantaneously rise to a heat flux everywhere in the body. Physically, it is expected that a change in the temperature gradient should be felt after some build-up or relaxation time, and that disturbances travel at finite velocity.

That theoretical paradox has been one of the principal incentives for the development of modern thermodynamic theories, because a macroscopic theory with finite speed of propagation also comes from the experimental observations of the second-sound and ballistic phonon propagation in some dielectric at low temperature [1, 17, 58–60, 84].

Several authors have dealt with the problem of infinite speed of propagation of thermal signals. The first were Cattaneo [12, 13] and Vernotte [80] which, basing their analyses on the kinetic theory, proposed a damped version of Eq. (1.2) to eliminate the paradox of infinite speed of propagation of thermal signals, and to comply with the experimental observations. They generalized Eq. (1.2) as in Eq. (1.6), namely, as

$$\tau_R \dot{\mathbf{q}} + \mathbf{q} = -\lambda \nabla T .$$

In fact, for the sake of a formal simplicity, if one assumes that the material functions are constant, then the use of Eqs. (1.6) and (1.8) allows to obtain the following *telegrapher's equation*

$$\tau_R \ddot{T} + \dot{T} = \frac{\lambda}{c_v} \nabla^2 T \tag{2.2}$$

instead of Eq. (2.1). From the mathematical point of view, it is a hyperbolic partial differential equation and allows the thermal pulses, or every high-frequency waves, to propagate through a rigid body in thermodynamic equilibrium with a finite speed given by Eq. (1.49), that is, $U_0 = \sqrt{\lambda / (\tau_R c_v)}$.

It is worth noticing that the MCV equation also received several criticisms. For example, a supposed violation of second law by Eq. (1.6) is known in the literature as Taitel Paradox [6, 70, 85]. However, as proved by Sharma [68], this is due to unrealistic initial time conditions. The paradox is removed if more realistic initial conditions are given, that is, one has to impose on the initial condition not only a positive profile for the temperature, but also a restriction on the heat flux **q**.

Indeed, without introducing a relaxational term one should observe that finite-speed thermal-disturbance propagations can be also obtained by assuming that the thermal conductivity λ in the FL depends on the temperature [8, 54]. However, the experiments show that at intermediate or high temperatures both in fluids, and in metals this dependence is not the one required to avoid the paradox of infinite speed of propagation.

In EIT, instead, finite speeds of propagation can be obtained in generalized sense, by observing that a correct estimation of the order of magnitude of the speed of propagation of the perturbations requires to compare the order of magnitude of the solution of the field equations with that of the error affecting the experimental data on which the theory is based [19]. Then, if in a finite interval of time the solution of the system of balance laws is bigger than the experimental error only in a compact domain, then such a solution is experimentally zero outside this domain. As a consequence, it has propagated with finite speed, because not all the points of the space have been reached in a finite time. This point of view complies with a classical result proved by Fichera in his celebrated defense of Fourier theory [34], whose meaning is that the measurements of temperature and heat flux are valid only within the limits of observation, which, in turn, depend on the sensitivities of the measuring instruments [25].

2.2 The Guyer-Krumhansl Equation and Phonon Hydrodynamics

The thermal properties of an insulating crystal at low temperatures can be described by means of the phonons in it [38]. Phonons, which are quasi-particles obeying the Bose-Einstein statistics [61], may interact among themselves and with lattice imperfections through two different types of processes:

1. Normal (N) processes, for which phonon momentum is conserved. These interactions are characterized by the relaxation time $\tau_N = 1/\nu_N$, with ν_N being the frequency of normal interactions.
2. Resistive (R) processes, for which phonon momentum is not conserved. These interactions are characterized by the relaxation time $\tau_R = 1/\nu_R$, with ν_R being the frequency of resistive interactions.

The second sound takes over whenever $\nu_N \gg \nu_R$. In this case, the MCV equation (1.6) accurately describes the second-sound phenomena. However, as shown by Guyer and Krumhansl in 1966, Eq. (1.6) is inadequate when $\nu_N \ll \nu_R$ since, in this case, it is not able to introduce nonlocal effects, which are important in the phonon flow [38, 39]. Therefore, starting from a linearized Boltzmann equation in the Callaway approximation, in three dimensions they established Eq. (1.17), namely,

$$\tau_R \dot{\mathbf{q}} + \mathbf{q} = -\lambda \nabla T + \ell_p^2 \left(\nabla^2 \mathbf{q} + 2 \nabla \nabla \cdot \mathbf{q} \right)$$

as the evolution equation for the heat flux. In that equation, the square of the phonon mean-free path $\ell_p^2 = \bar{v}^2 \tau_N \tau_R / 5$ determines the nonlocal term. Moreover, in Eq. (1.17) λ means the usual Ziman limit [87] for the bulk thermal conductivity, i.e., $\lambda = c_v \bar{v} \ell_p' / 3$, with $\ell_p' = \bar{v} \tau_R$. The collective hydrodynamic aspects of normal collisions have been recently stressed out in a microscopic kinetic-collective model of heat transport in Refs. [27, 28].

Macroscopic derivations of Eq. (1.17) can be also obtained in weakly nonlocal nonequilibrium thermodynamics [18, 20, 43, 50, 77]. For example, if in Eq. (1.41) nonlocal terms depending on the spatial derivatives of β are introduced as

$$\dot{\beta} = -\frac{1}{\tau_R} (\beta - T) + 3 \frac{\ell_p^2}{\tau_R} \nabla^2 \beta \tag{2.3}$$

straightforward calculations allow to recover Eq. (1.17) in the case of constant material functions.

The Guyer-Krumhansl (GK) equation (1.17) can accurately describe heat transport in several physical situations. In particular, at low-temperature regime, and when $\tau_N \ll \tau_R$, Eq. (1.17) still predicts that the speed of thermal pulses is Eq. (1.49) [43]. At high frequencies, and when the relaxation times above have same order of magnitude, a ballistic phonon propagation is observed, that is, in such a situation phonons travel through the whole crystal without suffering any collisions. In this case, the speed (1.49) becomes [43]

$$U_0 = \sqrt{\frac{\lambda}{c_v \tau_R} + \frac{3 \ell_p^2}{\tau_R \tau_N}}. \tag{2.4}$$

Furthermore, it is also able to describe the regime known as phonon hydrodynamics, where the term in $\ell_p^2 \nabla^2 \mathbf{q}$ plays a specially relevant role (analogous to the role of viscous terms in classical hydrodynamics). This describes several aspects of the size dependence of heat conductivity in nanosystems, as it will be shown in Chap. 3. This kind of heat-transfer equations are also found in heat propagation in turbulent superfluid helium [55, 56, 62, 63, 79].

Phonon hydrodynamics has been also deeply studied from the microscopic perspective in the framework of maximum-entropy formalism [3–5], dealing also with nonlinear contributions and dispersive effects in the dispersion relation for phonon velocities [45–48]. Other microscopic modelizations of phonon hydrodynamics may be found in [36, 41].

2.3 The Nonlinear MCV and GK Equations

From the theoretical point of view, the search for a generalization of the FL which allows to avoid the paradox of infinite speed of propagation of thermal disturbances is related to the problem of finding a suitable equation leading to a hyperbolic partial differential equation governing the evolution of the temperature.

As previously observed, the introduction of relaxational effects as in Eq. (1.6) allows to reach this goal. A great part of the works dealing with heat-waves propagation is focused on the analysis of the consequences of the dynamical behavior of the generalized heat-transport equation. This is a natural consequence of the important role played by relaxational terms in high-frequency nonequilibrium situations [44]. However, the same attention should be put on nonlinear terms which are also important in strong nonequilibrium situations, as heat conduction at nanoscale.

From the thermodynamic point of view, in Sect. 1.1 we already observed that the use of Eq. (1.6), for a rigid heat conductor, requires that the form of the entropy per unit volume s is no longer the local-equilibrium one $s_{eq}(u)$, but it should take the form (1.13) [43].

Indeed, if one admits that in nonequilibrium situations the entropy per unit volume is given by Eq. (1.13), then the use of the dynamical nonequilibrium temperature β allows to obtain a nonlinear generalization of the classical MCV equation.

In fact, if one assumes that the state space is spanned both by u and β, and by their first-order spatial derivatives ∇u and $\nabla \beta$ in view of a weakly nonlocal description, then it is possible to generalize Eq. (1.41) as

$$\dot{\beta} = f(u; \beta; \nabla u; \nabla \beta) \tag{2.5}$$

with f being a regular function of the indicated arguments, the form of which follows from thermodynamic restrictions. In particular, as a consequence of second law of thermodynamics [18, 21, 22], it follows that in Eq. (2.5)

$$f = f_{eq}(u; \beta) + \left(\frac{\lambda}{2s_\beta}\right)\left(\frac{\partial s_\beta}{\partial u}\right)\nabla\beta \cdot \nabla\beta \tag{2.6}$$

with $f_{eq}(u; \beta)$ being a regular function of the indicated arguments, and s_β being the nonequilibrium part of a generalized entropy density reading [22]

$$s = s_{eq}(u; \beta) - \left(\frac{s_\beta}{2}\right) \nabla\beta \cdot \nabla\beta. \tag{2.7}$$

Without losing the generality, it is possible to assume [22]

$$f_{eq}(u; \beta) = -\frac{1}{\tau_R}(\beta - T) \tag{2.8}$$

whereas the comparison between Eqs. (1.13) and (2.7) turns out Eq. (2.8) for s_β [22]. Therefore, along with these results, in a more general framework the dynamical nonequilibrium temperature β is ruled by Eq. (1.46), which in an explicit form is

$$\dot{\beta} = -\frac{1}{\tau_R}(\beta - T) - \left(\frac{\lambda}{Tc_v}\right)\nabla\beta \cdot \nabla\beta \tag{2.9}$$

and generalizes the MCV equation (1.6) as [22, 23]

$$\tau_R\dot{\mathbf{q}} + \mathbf{q} = -\lambda\nabla T + \left(\frac{2\tau_R}{Tc_v}\right)\nabla\mathbf{q} \cdot \mathbf{q}. \tag{2.10}$$

In contrast with Eq. (1.6), Eq. (2.10) explicitly introduces nonlinear terms, whose consequences may be pointed out by studying heat-propagation again. In fact, still assuming that all the material functions are constant, in the case of heat transport in a one-dimensional rigid body (x means the only one cartesian component), the coupling of Eqs. (1.8) and (2.10) allows to obtain the following hyperbolic partial differential equation [22]

$$\tau_R\ddot{T} + \dot{T} - \frac{2\tau_R q_0}{c_v T}\nabla_x\dot{T} = \frac{\lambda}{c_v}\nabla_x^2 T \tag{2.11}$$

with q_0 being the average value of the heat flux flowing through the system. That equation is more general than Eq. (2.2) and leads to

$$U^+ = U_0\left(\sqrt{\phi^2 + 1} - \phi\right) \tag{2.12}$$

for the speed of propagation of the pulses in the positive direction, namely, in the same direction of the heat flux, and to

$$U^- = U_0\left(\sqrt{\phi^2 + 1} + \phi\right) \tag{2.13}$$

in the opposite direction. In Eqs. (2.12) and (2.13) U_0 is given by Eq. (1.49), and $\phi = q_0/(U_0 T c_v)$.

In other words, Eq. (2.12) points out that a small heat pulse, moving in the same direction of the heat flow, will travel with a velocity which differs from the speed of propagation of pulses in opposite direction, given by Eq. (2.13). In particular, the theoretical model in Eq. (2.10) prescribes that the former speed is smaller than the latter one [22, 23]. A similar result has been obtained in Ref. [11] and in Ref. [26] by applying different approaches.

It is important to underline that Eqs. (2.12) and (2.13) must be understood as thermodynamic predictions of a relation between the speeds of thermal pulses in equilibrium, which give information on τ_R, and the speeds of thermal pulses under a heat flux.

In Sect. 2.2 we observed that the MCV equation is inadequate to describe heat transfer whenever the relaxation time of normal interactions between phonons is not negligible with respect to the resistive relaxation time. Therefore, in order to have a general model able to account for the different phonon contributions to the heat conduction in rigid solids, Guyer and Krumhansl derived Eq. (1.17).

Although being very refined, Eq. (1.17) is not able to account for nonlinear effects. Indeed, a macroscopic generalization of the GK equation incorporating nonlinear quadratic terms, analogous to the extra terms appearing in Eq. (2.10), may be obtained by considering the following evolution equation for β

$$\dot{\beta} = -\frac{1}{\tau_R}(\beta - T) - \frac{\lambda}{Tc_v}\nabla\beta \cdot \nabla\beta + 3\frac{\ell_p^2}{\tau_R}\nabla^2\beta. \tag{2.14}$$

Direct calculations, in fact, point out that Eq. (2.14) extends Eq. (1.17) to the nonlinear case as [23, 24]

$$\tau_R\dot{\mathbf{q}} + \mathbf{q} = -\lambda\nabla T + \left(\frac{2\tau_R}{Tc_v}\right)\nabla\mathbf{q} \cdot \mathbf{q} + \ell_p^2\left(\nabla^2\mathbf{q} + 2\nabla\nabla \cdot \mathbf{q}\right) \tag{2.15}$$

which can be further generalized in

$$\tau_R\dot{\mathbf{q}} + \mathbf{q} + \left(\mu\nabla\mathbf{q} + \mu'\nabla^t\mathbf{q}\right) = -\lambda\left(1 + \xi\mathbf{q} \cdot \mathbf{q}\right)\nabla T + \ell_p^2\left(\nabla^2\mathbf{q} + 2\nabla\nabla \cdot \mathbf{q}\right). \tag{2.16}$$

This represents a generalized nonlinear evolution equation for the heat flux which allows to consider, in a simple way, some aspects of memory, nonlocal, and nonlinear effects, besides deriving equations for the propagation of heat waves, which is a well-known topic in current nonequilibrium thermodynamics [32, 42, 43, 51, 57]. An equation like Eq. (2.16) has been proposed in the thermomass theory [10, 29–31, 76], as we will see in Sect. 2.4. In practical applications, some consequences of Eq. (2.14) will be pointed out in Chap. 5.

2.3.1 Characteristic Dimensionless Numbers for Heat Transport: Approximated Heat-Transport Equation

Every problem in thermodynamics (as well as, in general, in continuum mechanics) is always characterized by two types of unknown quantities: the intensive variables and the extensive ones. These variables are, indeed, unknown functions since, in a given thermodynamic system, they may vary both in time, and in space. For each of those unknown functions, suitable equations can be introduced, which are always of the balance-type in the case of basic extensive variables, of the evolution-type in the case of basic intensive variables, and in the form of constitutive equations for all non-basic unknown functions.

Although from the theoretical point of view the set-up of a thermodynamic problem is often quite easy, its solution may be, instead, very hard in practical applications. Therefore, to reduce the problem to a simpler level, in many cases in the model equations one may neglect one, or more terms. Indeed, each term, entering in a given equation which models a physical problem, can be interpreted as the cause which will produce variations in the unknown functions. Far from instability conditions, it is always possible to establish a strict relation between causes and effects, and therefore, when a term in the model equation (that is, the cause) is neglected, then the consequent effect will be neglected, too.

However, it is evident that the results one obtains in this way can not always be applied, since they hold only for the special case characterized by the simplifying assumptions that have been made, and once those assumptions break down, the derived conclusions are no longer valid. It is important, therefore, the establishment of suitable criteria to control a priori the validity of each assumption.

A logical criterion implies to recognize the measure of the relative importance of an effect with respect to the others. To this end and since one may compare only effects which are homogeneous in dimensions, it is useful to introduce some dimensionless numbers, which are often called characteristic numbers, too.

In the next we denote by a superscript $*$ the dimensionless unitary quantities, and by a superscript r the reference quantities, i.e., the standard values of the physical quantities at which the system at hand is running. We stress the fact that the reference quantities depend on the class of phenomena which the physical system undergoes, and they can be determined only within a well-established experimental framework.

Looking at the local balance of energy (1.8), we may observe that the relative importance of the left-hand side term with respect to the right-hand side one turns out a suitable measure of the unsteady state of u. To understand the conditions under which we may neglect the temporal rate of the internal energy (namely, in what situations we are allowed to consider the problem as a quasi steady-state process), taking into account the relation $du = c_v dT$, let us rewrite Eq. (1.8) as

$$\left(\frac{c_v^r T^r}{t^r}\right) c_v^* \dot{T}^* + \left(\frac{q^r}{L^r}\right) \nabla \cdot \mathbf{q}^* = 0 \Longrightarrow \left(\frac{1}{\mathrm{M}_q}\right) \dot{u}^* + \nabla \cdot \mathbf{q}^* = 0 \qquad (2.17)$$

with

$$t^r = \frac{c_v^r L^{r2}}{\lambda^r} \tag{2.18a}$$

$$M_q = \frac{q^r L^r}{\lambda^r T^r}. \tag{2.18b}$$

In practical applications, one may try to interpret T^r as the average temperature of the system, λ^r and c_v^r as the experimental values, respectively, of the thermal conductivity and of the specific heat at constant volume of the system at T^r, L^r as the longitudinal length of the system, and q^r as the longitudinal heat flow due to the difference in temperature through the ends of the system.

From Eqs. (2.17) and (2.18) one may conclude that for regular solutions of the field equations, i.e., for finite values of $\dot{u}$, it is possible to neglect the term in $\dot{u}$ whenever $M_q \gg 1$. Forcing the usual fluid-dynamic nomenclature, we refer to the dimensionless number (2.18b) as the Mach number for the heat flow. Indeed, it is possible to observe that, since in fluid dynamics the Mach number (M) is given by the ratio between the body speed and the speed of sound (assumed as the reference speed), analogously, in our model M_q is given by the ratio between the actual heat flux in the system (i.e., q^r) and a reference heat flux (i.e., $\lambda^r T^r/L^r$). Moreover, since in fluid dynamics M represents the percent rate, per unit time, of the mass density along a given particle stream-line (and therefore it is a measure of the compressibility of the fluid), in Eq. (1.8) M_q will indicate the temporal rate of u.

From the practical point of view one may try to check a critical value of M_q representing the threshold between the steady-state problem and the unsteady one. Experimental situations could complement and test the validity of this theoretical proposal. However, let us underline that the possibility of interpreting a characteristic number as a measure of the relative importance of an effect with respect to another one (and all the arising consequences) is always conditioned by the appropriate determination of the reference quantities.

Let us concentrate now our attention on the nonlinear equation (2.15), describing the evolution of the heat flux. According with the conclusions above, when $M_q \gg 1$ the heat-flux vector becomes solenoidal, and Eq. (2.15) can be rewritten as

$$\left(\frac{q^r}{t^r}\right)\dot{\mathbf{q}}^* + \left(\frac{q^r}{\tau_R^r}\right)\frac{\mathbf{q}^*}{\tau_R^*} - \left(\frac{2q^{r2}}{c_v^r T^r L^r}\right)\frac{\nabla \mathbf{q}^* \cdot \mathbf{q}^*}{c_v^* T^*}$$

$$- \left(\frac{q^r \ell_p^{r2}}{\tau_R^r L^{r2}}\right)\frac{\ell_p^{*2}}{\tau_R^*}\nabla^2 \mathbf{q}^* + \left(\frac{\lambda^r T^r}{\tau_R^r L^r}\right)\frac{\lambda^*}{\tau_R^*}\nabla T^* = \mathbf{0} \Longrightarrow$$

$$\frac{\dot{\mathbf{q}}^*}{St_q} + \left(\frac{Re_q}{Kn_q^2}\right)\frac{\mathbf{q}^*}{\tau_R^*} - \frac{\nabla \mathbf{q}^* \cdot \mathbf{q}^*}{c_v^* T^*} - \left(\frac{1}{Re_q}\right)\frac{\ell_p^{*2}}{\tau_R^*}\nabla^2 \mathbf{q}^* + \left(\frac{1}{Fr_q}\right)\frac{\lambda^*}{\tau_R^*}\nabla T^* = \mathbf{0}$$

$$\tag{2.19}$$

wherein we have introduced the following characteristic numbers

$$St_q = 2\frac{q^r t^r}{c_v^r T^r L^r} \tag{2.20a}$$

$$Re_q = 2\frac{q^r \tau_R^r L^r}{\ell_p^{r2} c_v^r T^r} \tag{2.20b}$$

$$Kn_q = \frac{\ell_p^r}{L^r} \tag{2.20c}$$

$$Fr_q = M_q \, Re_q \, Kn_q^2 = 2\frac{q^{r2} \tau_R^r}{\lambda^r c_v^r T^{r2}} \tag{2.20d}$$

which, along with the same observations as above, we call, respectively, thermal Strouhal number, thermal Reynolds number, thermal Knudsen number, and thermal Froude number. Under the same identifications as above for the reference quantities, it is also possible to identify ℓ_p^r and τ_R^r, respectively, as the mean-free path of phonons and the resistive relaxation time at the average temperature T^r. The reference time t^r used in Eq. (2.19), in principle, would be different from that used in Eq. (2.17). However, if one uses for it the same reference time defined in Eq. (2.18a), then from Eq. (2.20a) it follows that $St_q = 2 M_q$: if the temporal rate of the internal energy can be neglected in its evolution equation, then in the evolution equation for the heat flux the temporal rate can be neglected, too.

Equation (2.19) points out that, whenever $Re_q / Kn_q^2 \ll 1$, it is possible to neglect the heat flux with respect to its temporal and spatial variations in Eq. (2.15). In this condition, which is frequently recovered at nanometric scale since the characteristic dimension of the system is smaller than the phonon mean-free path (that is, $Kn_q \gg 1$ in usual practical applications) [23], Eq. (2.15) becomes

$$\tau_R \dot{\mathbf{q}} - \left(\frac{2\tau_R}{Tc_v}\right) \nabla \mathbf{q} \cdot \mathbf{q} = -\lambda \nabla T + \ell_p^2 \nabla^2 \mathbf{q} \tag{2.21}$$

which is very similar to the Navier-Stokes equation describing the motion of an incompressible viscous fluid in the absence of external force, and gives a better understanding of the strict relation between the hydrodynamic quantities and the thermal ones. Moreover, Eq. (2.21) allows to relate the term in $\tau_R/ (Tc_v) \nabla \mathbf{q} \cdot \mathbf{q}$ with the convective phenomena occurring during the evolution of the heat flux, the term in $\ell_p^2 \nabla^2 \mathbf{q}$ with the non-Fourier diffusive phenomena, and the term in $\lambda \nabla T$ with the Fourier heat conduction generated by the inhomogeneities in the thermal field. Note that the possibility of having both a convective contribution and a diffusive one, should not be surprising. In fact, in the framework of EIT [43] the heat flux has its own evolution equation, which, under suitable hypotheses on the constitutive equations for the flux and the production of heat flux, can be put in the form

$$\dot{\mathbf{q}} + \frac{\mathbf{q}}{\tau_R} + \nabla \cdot (\mathbf{Q}_\text{c} + \mathbf{Q}_\text{d}) + \nabla \cdot \left(\frac{\lambda T}{\tau_R} \mathbf{I} \right) = \mathbf{0} \qquad (2.22)$$

wherein $\mathbf{Q}_\text{c}$ and $\mathbf{Q}_\text{d}$ are second-order tensors representing the convective part of the heat flux and its diffusive non-Fourier part, respectively, $(\lambda T/\tau_R)\,\mathbf{I}$ represents the Fourier contribution to the flux of heat flux, and the term $-\mathbf{q}/\tau_R$ is the production of heat flux.

The comparison of Eqs. (2.19) and (2.22) allows us to claim that in the latter equation the thermal Strouhal number (2.20a) gives information about the relative importance of the non-stationary term with respect to the convective one, whereas the thermal Froude number (2.20d) accounts for the relative importance of the Fourier diffusive term with respect to the convective one, in analogy with the physical meanings of hydrodynamic Strouhal (St) and Froude (Fr) numbers. However, in contrast with the formal analogy between the definitions of the thermal Mach number (2.18b) and the hydrodynamic Mach number, in classical hydrodynamics St and Fr have definitions which are not formally the same of those in Eqs. (2.20a) and (2.20d), although they play an analogous role. Indeed, in hydrodynamics one has $\text{St} = t^r v^r / L^r$, and $\text{Fr} = v^{r2} / (L^r \bar{g}^r)$, being v^r the reference fluid speed and $\bar{g}^r$ a reference acceleration, which is usually taken as the gravity acceleration, but $q^r / c_v^r T^r$ has dimensions of a speed, i.e., units of m/s.

Finally, we observe that in Eq. (2.19), the thermal Reynolds number (2.20b) points out, instead, the relative importance of the diffusive non-Fourier term with respect to the convective one. Its definition is similar to the definition of hydrodynamic Reynolds number (Re), i.e., $\text{Re} = v^r L^r / v^r$, being v^r the reference kinematic viscosity. Thus, in our hydrodynamic analogy, the ratio $2\tau_R / \left(\ell_p^2 T c_v \right)$ plays the same role of the kinematic viscosity v in classical hydrodynamics. Further information about the relative importance between other effects may be enlightened by obtaining suitable ratio between these characteristic numbers.

2.4 Other Generalized Heat-Transport Equations

2.4.1 The Dual-Phase-Lag Model

The advancement of ultrashort pulsed lasers inspired ultra-fast heat transport occurring in times comparable to the mean-free times of energy carriers [73, 75, 76]. The individual behaviors of those carriers become pronounced owing to their scant number of collisions in such short times. As a consequence, the traditional concepts in heat transport based on the averaged behaviors over many collisions can no longer be applied. Evolving from the Fourier law (FL) describing the quasi-stationary and reversible transition of thermodynamic states, in the dual-phase-lag model [72, 74, 75] the following constitutive equation (which provides a convenient approach in describing the ultra-fast physical response) is introduced

$$\mathbf{q}\left(\mathbf{r}, t + \tau_q\right) = -\lambda \nabla T \left(\mathbf{r}, t + \tau_T\right) \Rightarrow$$

$$\mathbf{q}\left(\mathbf{r}, t\right) + \tau_q \dot{\mathbf{q}}\left(\mathbf{r}, t\right) + \frac{\tau_q^2}{2} \ddot{\mathbf{q}}\left(\mathbf{r}, t\right) \approx -\lambda \nabla T \left(\mathbf{r}, t\right) - \tau_T \nabla \dot{T}\left(\mathbf{r}, t\right) + \frac{\tau_T^2}{2} \nabla \ddot{T}\left(\mathbf{r}, t\right)$$
(2.23)

where τ_T and τ_q are two phase lags due to the delayed response during the ultra-fast transient. In particular, the phase of the heat flux τ_q captures the small-scale response in time, and the phase lag of the temperature gradient τ_T captures the small-scale response in space.

As it can be seen from Eq. (2.23), the lagging response significantly deviates from FL due to involvement of the high-order derivatives with respect to time. Moreover, only when $\tau_T = 0$, and the terms proportional to τ_q^2 can be neglected, Eq. (2.23) reduces to the thermal-wave equation (1.6) proposed by Cattaneo [12, 13] and Vernotte [80]. In correlation to the existing microscopic models, in addition, the lagging behavior described by Eq. (2.23) absorbs electron-phonon coupling in metals, umklapp (temporary momentum loss) and normal relaxations in phonon scattering, and additional relaxation of internal energy in the same framework of thermal lagging [76].

The combination of the divergence of Eq. (2.23) with the energy-balance equation in absence of source terms (1.8) leads to

$$\nabla^2 T + \tau_T \nabla^2 \dot{T} + \frac{\tau_T^2}{2} \nabla^2 \ddot{T} = \left(\frac{c_v}{\lambda}\right) \dot{T} + \left(\frac{c_v \tau_q}{\lambda}\right) \ddot{T} + \left(\frac{c_v \tau_q^2}{2\lambda}\right) \dddot{T}$$
(2.24)

which completely alters the fundamental characteristics of Fourier diffusion due to the presence of the mixed-derivative terms and the high-order derivatives of temperature with respect to time. When the effects related to τ_T^2 in Eq. (2.23) are neglected, Eq. (2.24) is able to describe the so-called T waves, whose speed U_T reads

$$U_T = \sqrt{\frac{2 c_v \tau_T}{\lambda \tau_q^2}}.$$
(2.25)

Whether a T wave is slower than the Maxwell-Cattaneo-Vernotte (MCV) wave (1.49), or not, depends on the ratio τ_T / τ_q since [75]

$$\frac{U_T}{U_0} = \sqrt{2 \left(\frac{\tau_T}{\tau_q}\right)}.$$
(2.26)

For example, for heat propagation in superfluid helium at very low temperatures one has $U_T < U_0$ since $\tau_T < \tau_q$. For femtosecond-laser heating on metal films, instead, $\tau_T > \tau_q$ and the T waves move faster that the MCV waves [75].

The effects of τ_T^2 are recently used to describe transport phenomena in biological systems with multiple energy carriers [76].

2.4.2   The Thermomass Theory

The Thermomass (TM) theory [10, 29–31] provides another example of heat-transport equation showing the nonlocal behavior with thermal lagging in transporting heat. In TM theory the heat transport is due to the motion of a gas-like collection of heat carriers, characterized by an effective mass density and flowing through the medium due to a thermomass-pressure gradient. This collection is made by quasi-particles of heat carriers, called thermons, which are representative of the vibrations of the molecules generated by heating the conductor and whose mass may be calculated from the Einstein's mass-energy duality. For crystals, the thermomass gas is just the phonon gas, for pure metals it will be attached on the electron gas, and for semi-metals it will be constituted by both these different gases [10, 29–31].

In TM theory the evolution equation for the heat flux reads [76]

$$\tau_{tm}\dot{\mathbf{q}} - c_v\dot{T}\mathbf{L} + \nabla\mathbf{q}\cdot\mathbf{L} + \lambda\left(1 - M_{tm}^2\right)\nabla T + \mathbf{q} = \mathbf{0} \qquad (2.27)$$

wherein

$$\tau_{tm} = \frac{\lambda\rho}{2\gamma c_v^2 T}$$

is the relaxation time in the TM theory [29, 64, 83] with γ being the Grüneisen constant, and

$$\mathbf{L} = \frac{\mathbf{q}\lambda\rho}{2\gamma c_v\left(c_v T\right)^2}$$

denotes a characteristic-length vector [64, 83] which characterizes the strength of the non-Fourier effects introduced by Eq. (2.27) [64, 83] and it should not be confused with the mean-free path of the thermons. In practical applications and for conceivable values of $\mathbf{q}$, $\mathbf{L}$ attains values which are always much smaller than those of the thermons mean-free path. Moreover, in Eq. (2.27)

$$M_{tm} = \frac{q\sqrt{\rho}}{c_v T\sqrt{2\gamma c_v T}} \qquad (2.28)$$

stands for a dimensionless number which is called thermal Mach number of the drift velocity relative to the thermal-wave speed in the heat-carrier collection, and should be not confused with the thermal Mach number introduced in Eq. (2.18b). By means of it, Eq. (2.27) introduces an effective nonlinear thermal conductivity, namely,

$$\lambda_{\text{eff,nl}} = \lambda \left(1 - M_{\text{tm}}^2\right) \tag{2.29}$$

which plays a relevant role when the thermons mean-free path is larger than the characteristic size of the system, i.e., at nanoscale. In such a case, in fact, the spatial derivatives of the heat flux may be neglected with respect to the flux itself so that, in steady states, Eq. (2.27) reduces to

$$\mathbf{q} = -\lambda_{\text{eff,nl}} \nabla T.$$

Since second law of thermodynamics requires $\lambda_{\text{eff,nl}} \geq 0$, from Eqs. (2.28) and (2.29) it follows that the modulus of the local heat flux has to fulfill the following relation

$$q \leq c_v T \sqrt{\frac{2\gamma c_v T}{\rho}} \tag{2.30}$$

which implies an upper bound for the heat flux for increasing temperature gradient [65].

The experimental results in silicon nanowires, for a difference of temperature $\Delta T = 100\,\text{K}$ confirm the existence of this upper bound [82]. Such a phenomenology, well-known in nonlinear heat conduction, is referred to as the presence of "flux limiters" [43, 66]. Flux limiters are a direct consequence of the finite speed of the thermal perturbations. In fact, for a given energy density, the heat flux can not reach arbitrarily high values, but it has to be bounded by a maximum saturation value of the order of the energy density times the maximum speed. Typical situations of flux limiters arise, for instance, in radiative heat transfer, or in plasma physics [2, 53].

Beside to describe relaxational effects, Eq. (2.27) incorporates information on the characteristic length of the system (i.e., nonlocal effects) and accounts for nonlinear phenomena. In the linear case (i.e., when the terms containing the quantities $\dot{T}\mathbf{L}$ and $\nabla\mathbf{q}\cdot\mathbf{L}$ are negligible, as well as when $M_{\text{tm}}^2 \ll 1$), Eq. (2.27) takes the same form of the MCV equation, even though, from the theoretical point of view, the relaxation time of TM theory is two orders of magnitude higher than that of the MCV's theory and, in silicon films, it predicts a slower response to the thermal perturbations [82].

2.4.3   Anisotropic Heat Transport

The dynamical temperature β does not constitute the sole tool to obtain nonlinear extensions of the GK equation. For instance, Lebon et al. [50] developed a continuum model of weakly nonlocal and nonlinear heat transport, resulting, in the linear case, in Eq. (1.17). In order to keep down the calculations to a reasonable level, these authors considered the isotropic case only. Notwithstanding, the crystals in which nonlocal heat conduction takes place, such as Sodium Fluoride, Bismuth or Silicon, are highly anisotropic so that a generalization of the model considered in [50], which encompasses anisotropic systems, seems to be necessary. By extending their approach to anisotropic situations, in Ref. [71] Triani and Cimmelli obtained an evolution equation for the heat flux which in components reads

$$\tau_{IJ}\dot{q}_{J}+q_{I} = -\tau_{IJ}\left(\frac{\lambda}{\tau_{R}} + \frac{1}{2c_{v}T^{2}}q^{2}\right)T_{,J} + \frac{2}{c_{v}T}\tau_{IJ}q_{K}q_{<K,J>} + \frac{9}{5}\frac{\lambda\tau_{N}}{c_{v}}\tau_{IJ}q_{L,LJ}$$

$$+ \frac{1}{c_{v}T}\tau_{IJ}q_{J}q_{<K,K>} - \frac{1}{c_{v}T^{2}}T_{,K}\tau_{IJ}q_{J}q_{K} + 2\tau_{IJ}\frac{\partial L_{JKMH}}{\partial T}T_{,K} < q_{M,H} >$$

$$+ 2\tau_{IJ}\frac{\partial L_{JKNH}}{\partial q_{M}}q_{M,K} < q_{N,H} > +\tau_{IJ}L_{JKNH}q_{N,HK} + \tau_{IJ}L_{JKNH}q_{H,NK}. \qquad (2.31)$$

In this equation, which represents an extension to anisotropic systems of Eq. (2.15), the symbol $F_{,J}$ represents the partial derivative of F with respect to the Cartesian coordinate X_{K}, $F_{<N,H>}$ denotes the symmetric part of the function $F_{N,H}$, and the Einstein's convention of summation over repeated indices has been adopted. In Eq. (2.31), a fundamental role is played by the matrix of the relaxation times τ_{IJ}, which represents the relaxational properties of the crystal along different directions, and by the tensor of the mean-free paths, which reflects the mobility properties of the phonons along different directions. The principle of material frame indifference [25] imposes severe restrictions on the form of these tensorial material functions [71]. If

$$L_{IKJH} = \frac{9}{5}\frac{\lambda\tau_{N}}{c_{v}\tau_{R}}\delta_{HI}\delta_{KJ}$$

and

$$\tau_{IJ}^{-1} = \frac{1}{\tau_{R}}\delta_{IJ}$$

it is possible to recover the following generalized GK equation for isotropic systems

$$\dot{q}_{I} + \frac{q_{I}}{\tau_{R}} = -\left(\lambda + \frac{1}{c_{v}T^{2}}q^{2}\right)T_{,I} + \frac{2}{c_{v}T}q_{K}q_{K,I} + \frac{9}{5}\frac{\lambda\tau_{N}}{c_{v}\tau_{R}}\left(q_{I,KK} + 2q_{K,KI}\right). \qquad (2.32)$$

Finally, for vanishing τ_N the equation above reduces to the further nonlinear MCV equation

$$\dot{q}_I + \frac{q_I}{\tau_R} = -\left(\lambda + \frac{1}{c_v T^2}q^2\right)T_{,I} + \frac{2}{c_v T}q_K q_{K,I}. \qquad (2.33)$$

As the TM equation (2.28), the last two equations show the presence of a genuinely nonlinear heat conductivity

$$\lambda_{\mathrm{nl}} = \left(\lambda + \frac{1}{c_v T^2}q^2\right) \qquad (2.34)$$

but, in contrast with it, the nonlinear term depending on the heat flux is positive. However, analogously to the case of TM theory, λ_{nl} cannot assume arbitrarily high values due to the presence of flux limiters [65].

2.4.4 Two-Population Ballistic-Diffusive Model

Up to now, we have considered phonons as a single population, but experiencing different kinds of collisions. Another extreme possibility, close to microscopic theories, would be to consider the phonons of each given frequency ω_p as a population on their own. An intermediate position [9, 14, 15, 37, 52, 86] rests on the assumption that two types of phonon populations may coexist: diffusive and ballistic phonons. In more details, diffusive phonons undergo multiple collisions within the core of the system, and ballistic phonons, originating at the boundaries of the system, experience mainly collisions with the walls. This model is called ballistic-diffusion (BD) model [49], and allows for a more flexible description of the heat transfer at nanoscale than the single-population models considered above.

On a purely macroscopic approach, in the BD model both the internal energy per unit volume u, and the local heat flux $\mathbf{q}$ are split into a ballistic part (bal) and a diffusive (dif) one in such a way

$$u = u_{\mathrm{bal}} + u_{\mathrm{dif}} \qquad \mathbf{q} = \mathbf{q}_{\mathrm{bal}} + \mathbf{q}_{\mathrm{dif}}. \qquad (2.35)$$

Owing to that decomposition, and according with EIT [43] the state variables are selected as follows:

- The couple $(u_{\mathrm{bal}}, \mathbf{q}_{\mathrm{bal}})$ accounts for the ballistic behavior of the heat carriers.
- The couple $(u_{\mathrm{dif}}, \mathbf{q}_{\mathrm{dif}})$ accounts for the diffusive behavior of the heat carriers.

The behaviors of u_{bal} and u_{dif} are provided by the following classical balance laws [49]:

$$\dot{u}_{\mathrm{bal}} = -\nabla \cdot \mathbf{q}_{\mathrm{bal}} + r_{\mathrm{bal}} \qquad (2.36a)$$

$$\dot{u}_{\text{dif}} = -\nabla \cdot \mathbf{q}_{\text{dif}}^{(e)} + r_{\text{dif}} \tag{2.36b}$$

wherein r_{bal} and r_{dif} denote, respectively, the source terms of the ballistic population and of the diffusive one. These quantities describe the energy exchange (per unit volume and time) between both phonon populations. It is easy to see that the summation of Eqs. (2.36a) and (2.36b) turns out the well-known balance law for u, which reduces to Eq. (1.8) whenever $r_{\text{bal}} = -r_{\text{dif}}$ in Eq. (2.36).

The behaviors of $\mathbf{q}_{\text{bal}}$ and $\mathbf{q}_{\text{dif}}$, instead, are given by [49]:

$$\tau_{\text{bal}}\dot{\mathbf{q}}_{\text{bal}} + \mathbf{q}_{\text{bal}} = -\lambda_{\text{bal}}\nabla T + \ell_{\text{bal}}^2 \left(\nabla^2 \mathbf{q}_{\text{bal}} + 2\nabla\nabla \cdot \mathbf{q}_{\text{bal}} \right) \tag{2.37a}$$

$$\tau_{\text{dif}}\dot{\mathbf{q}}_{\text{dif}} + \mathbf{q}_{\text{dif}} = -\lambda_{\text{dif}}\nabla T \tag{2.37b}$$

wherein τ_{bal} and τ_{dif} are the relaxation times of two phonon populations, λ_{bal} and λ_{dif} are their thermal conductivity, respectively, and ℓ_{bal} is the mean-free path of ballistic phonons. The relaxation times, the thermal conductivities and the phonon mean-free paths of two populations are not independent, but they are such that

$$\lambda_{\text{bal}} = \frac{1}{3}c_v v_{\text{bal}}^2 \tau_{\text{bal}} \qquad \lambda_{\text{dif}} = \frac{1}{3}c_v v_{\text{dif}}^2 \tau_{\text{dif}} \tag{2.38}$$

wherein $v_{\text{bal}} = \ell_{\text{bal}}/\tau_{\text{bal}}$ and $v_{\text{dif}} = \ell_{\text{dif}}/\tau_{\text{dif}}$ are the mean velocity of the ballistic and diffusive phonons, respectively, with ℓ_{dif} being the mean-free path of diffusive phonons [49]. From a microscopic perspective, one possibility is to consider the diffusive relaxation time τ_{dif} as that of resistive phonon collisions (i.e., the relaxation time τ_R), and the ballistic relaxation time τ_{bal} as that corresponding to the momentum-conserving collisions (i.e., the relaxation time τ_N) which lead to collective hydrodynamic effects. In this interpretation, the coupling of Eq. (2.37) leads to the GK equation (1.17).

At the very end, let us also observe that in the BD model, ballistic and diffusive phonons are allowed to have also different temperatures, namely, it is possible to define the following quasi-temperatures

$$T_{\text{bal}} = \frac{u_{\text{bal}}}{c_{v,\text{bal}}} \tag{2.39a}$$

$$T_{\text{dif}} = \frac{u_{\text{dif}}}{c_{v,\text{dif}}} \tag{2.39b}$$

wherein $c_{v,\text{bal}}$ and $c_{v,\text{dif}}$ denote the (positive) heat capacities per unit volume of the two populations. Admitting that $c_{v,\text{bal}} = c_{v,\text{dif}} = c_v$, and defining the quasi-temperature $T = u/c_v$, it is verified that $T = T_{\text{bal}} + T_{\text{dif}}$.

2.4.5   *Effective Medium Approach*

One could also mention, for the sake of a more exhaustive view, the so-called effective medium approach, which is sometimes used in the context of superlattices, nanofluids and nanoporous systems, and which will be illustrated in Chap. 4. This approach studies the thermal conductivity of systems composed of a heat conducting matrix (the thermal conductivity of which is λ_m), with embedded particles of thermal conductivity λ_p. It assumes that the classical expressions, derived by using the classical FL, are still formally valid, but with λ_m and λ_p suitably modified to take into account that the phonon mean-free path in the matrix may be comparable to the interparticle separation, or to the radius of the embedded particles. When these connections, as well as the thermal resistance between the matrix and the particles, are taken into account, the results describe qualitatively the main observed trends. Such connections, for instance, could be those mentioned in Eqs. (1.26)–(1.28). Indeed, this approach is not so basic as the ones previously mentioned in this section, because it borrows some of their results, combining them to classical expression, but it is helpful for dealing in a first approximation with the complexity in superlattices, nanocomposites, nanofluids and other analogous systems.

References

1. Ackerman, C.C., Bertman, B., Fairbank, H.A., Guyer, R.A.: Second sound in solid helium. Phys. Rev. Lett. **16**, 789–791 (1966)
2. Anile, A.M., Pennisi, S., Sammartino, M.: A thermodynamical approach to Eddington factors. J. Math. Phys. **32**, 544–550 (1991)
3. Banach, Z., Larecki, W.: Nine-moment phonon hydrodynamics based on the modified Grad-type approach: formulation. J. Phys. A: Math. Gen. **37**, 9805–9829 (2004)
4. Banach, Z., Larecki, W.: Nine-moment phonon hydrodynamics based on the maximum-entropy closure: one-dimensional flow. J. Phys. A: Math. Gen. **38**, 8781–8802 (2005)
5. Banach, Z., Larecki, W.: Chapman-Enskog method for a phonon gas with finite heat flux. J. Phys. A: Math. Gen. **41**, 375502 (18 pp.) (2008)
6. Barletta, A., Zanchini, E.: Unsteady heat conduction by internal-energy waves in solids. Phys. Rev. B **55**, 14208 (5 pp.) (1997)
7. Boltzmann, L.: Leçons sur la Théorie des Gaz. Gauthier-Villars, Paris (1902)
8. Bubnov, V.A.: Wave concepts in the theory of heat. Int. J. Heat Mass Transf. **19**, 175–184 (1976)
9. Cahill, D.C., Ford, W.K., Goodson, K.E., Mahan, G.D., Majumdar, A., Maris, H.J., Merlin, R., Phillpot, S.R.: Nanoscale thermal transport. J. Appl. Phys. **93**, 793–818 (2003)
10. Cao, B.-Y., Guo, Z.-Y.: Equation of motion of a phonon gas and non-Fourier heat conduction. J. Appl. Phys. **102**, 053503 (6 pp.) (2007)
11. Casas-Vázquez, J., Jou, D.: Nonequilibrium equations of state and thermal waves. Acta Physiol. Hung. **66**, 99–115 (1989)
12. Cattaneo, C.: Sulla conduzione del calore. Atti Sem. Mat. Fis. Univ. Modena **3**, 83–101 (1948)
13. Cattaneo, C.: Sur une forme de l'équation de la chaleur éliminant le paradoxe d'une propagation instantanée. C. R. Acad. Sc. **247**, 431–433 (1958)
14. Chen, G.: Ballistic-diffusion equations for transient heat conduction from nano to macroscales. J. Heat Transf. - T. ASME **124**, 320–328 (2001)

15. Chen, G.: Ballistic-diffusive heat-conduction equations. Phys. Rev. Lett. **86**, 2297–2300 (2001)
16. Chen, G.: Nanoscale Energy Transport and Conversion - A Parallel Treatment of Electrons, Molecules, Phonons, and Photons. Oxford University Press, Oxford (2005)
17. Chester, M.: Second sound in solids. Phys. Rev. **131**, 2013–2015 (1972)
18. Cimmelli, V.A.: An extension of Liu procedure in weakly nonlocal thermodynamics. J. Math. Phys. **48**, 113510 (13 pp.) (2007)
19. Cimmelli, V.A.: Different thermodynamic theories and different heat conduction laws. J. Non-Equilib. Thermodyn. **34**, 299–333 (2009)
20. Cimmelli, V.A., Frischmuth, K.: Gradient generalization to the extended thermodynamic approach and diffusive-hyperbolic heat conduction. Physica B **400**, 257–265 (2007)
21. Cimmelli, V.A., Sellitto, A., Triani, V.: A new thermodynamic framework for second-grade Korteweg-type viscous fluids. J. Math. Phys. **50**, 053101 (16 pp.) (2009)
22. Cimmelli, V.A., Sellitto, A., Jou, D.: Nonlocal effects and second sound in a nonequilibrium steady state. Phys. Rev. B **79**, 014303 (13 pp.) (2009)
23. Cimmelli, V.A., Sellitto, A., Jou, D.: Nonequilibrium temperatures, heat waves, and nonlinear heat transport equations. Phys. Rev. B **81**, 054301 (9 pp.) (2010)
24. Cimmelli, V.A., Sellitto, A., Jou, D.: Nonlinear evolution and stability of the heat flow in nanosystems: beyond linear phonon hydrodynamics. Phys. Rev. B **82**, 184302 (9 pp.) (2010)
25. Cimmelli, V.A., Jou, D., Ruggeri, T., Ván, P.: Entropy principle and recent results in non-equilibrium theories. Entropy **16**, 1756–1807 (2014)
26. Coleman, B.D., Fabrizio, M., Owen, D.R.: On the thermodynamics of second sound in dielectric crystals. Arch. Ration. Mech. Anal. **80**, 135–158 (1982)
27. De Tomas, C., Cantarero, A., Lopeandia, A.F., Alvarez, F.X.: From kinetic to collective behavior in thermal transport on semiconductors and semiconductor nanostructures. J. Appl. Phys. **115**, 164314 (2014)
28. De Tomas, C., Cantarero, A., Lopeandia, A.F., Alvarez, F.X.: Thermal conductivity of group-IV semiconductors from a kinetic-collective model. Proc R. Soc. A **470**, 20140371 (12 pp.) (2014)
29. Dong, Y., Cao, B.-Y., Guo, Z.-Y.: Generalized heat conduction laws based on thermomass theory and phonon hydrodynamics. J. Appl. Phys. **110**, 063504 (6 pp.) (2011)
30. Dong, Y., Cao, B.-Y., Guo, Z.-Y.: General expression for entropy production in transport processes based on the thermomass model. Phys. Rev. E **85**, 061107 (8 pp.) (2012)
31. Dong, Y., Cao, B.-Y., Guo, Z.-Y.: Temperature in nonequilibrium states and non-Fourier heat conduction. Phys. Rev. E **87**, 032150 (8 pp.) (2013)
32. Dreyer, W., Struchtrup, H.: Heat pulse experiments revisited. Contin. Mech. Thermodyn. **5**, 3–50 (1993)
33. Ferry, D.K., Goodnick, S.M.: Transport in Nanostructures, 2nd edn. Cambridge University Press, Cambridge (2009)
34. Fichera, G.: Is the Fourier theory of heat propagation paradoxical? Rend. Circ. Mat. Palermo **41**, 5–28 (1992)
35. Fourier, J.: The Analytical Theory of Heat. Cambridge University Press, Cambridge (1878)
36. Fryer, M.J., Struchtrup, H.: Moment model and boundary conditions for energy transport in the phonon gas. Contin. Mech. Thermodyn. **26**, 593–618 (2014)
37. Grmela, M., Lebon, G., Dauby, P.C., Bousmina, M.: Ballistic-diffusive heat conduction at nanoscale: GENERIC approach. Phys. Lett. A **339**, 237–245 (2005)
38. Guyer, R.A., Krumhansl, J.A.: Solution of the linearized phonon Boltzmann equation. Phys. Rev. **148**, 766–778 (1966)
39. Guyer, R.A., Krumhansl, J.A.: Thermal conductivity, second sound and phonon hydrodynamic phenomena in nonmetallic crystals. Phys. Rev. **148**, 778–788 (1966)
40. Hill, T.L.: Thermodynamics of Small Systems. Dover, New York (1994)
41. Jiaung, W.-S., Ho, J.-R.: Lattice-Boltzmann modeling of phonon hydrodynamics. Phys. Rev. E **6**, 066710 (13 pp.) (2008)
42. Joseph, D.D., Preziosi, L.: Heat waves. Rev. Mod. Phys. **61**, 41–73 (1989)

43. Jou, D., Casas-Vázquez, J., Lebon, G.: Extended Irreversible Thermodynamics, 4th revised edn. Springer, Berlin (2010)
44. Jou, D., Sellitto, A., Alvarez, F.X.: Heat waves and phonon-wall collisions in nanowires. Proc. R. Soc. A **467**, 2520–2533 (2011)
45. Larecki, W., Banach, Z.: Consistency of the phenomenological theories of wave-type heat transport with the hydrodynamics of a phonon gas. J. Phys. A: Math. Theor. **43**, 385501 (24 pp.) (2010)
46. Larecki, W., Banach, Z.: Influence of nonlinearity of the phonon dispersion relation in wave velocities in the four-moment maximum entropy phonon hydrodynamics. Physica D **266**, 65–79 (2014)
47. Larecki, W., Piekarski, S.: Phonon gas hydrodynamics based on the maximum entropy principle and the extended field theory of a rigid conductor of heat. Arch. Mech. **46**, 163–190 (1991)
48. Larecki, W., Piekarski, S.: Symmetric conservative form of low-temperature phonon gas hydrodynamics I. - Kinetic aspect of the theory. Nuovo Cimento D **13**, 31–53 (1991)
49. Lebon, G.: Heat conduction at micro and nanoscales: a review through the prism of extended irreversible thermodynamics. J. Non-Equilib. Thermodyn. **39**, 35–59 (2014)
50. Lebon, G., Jou, D., Casas-Vázquez, J., Muschik, W.: Weakly nonlocal and nonlinear heat transport in rigid solids. J. Non-Equilib. Thermodyn. **23**, 176–191 (1998)
51. Lebon, G., Jou, D., Casas-Vázquez, J.: Understanding Nonequilibrium Thermodynamics. Springer, Berlin (2008)
52. Lebon, G., Machrafi, H., Grmela, M., Dubois, C.: An extended thermodynamic model of transient heat conduction at sub-continuum scales. Proc. R. Soc. A **467**, 3241–3256 (2011)
53. Levermore, C.D.: Relating Eddington factors to flux limiters. J. Quant. Spectrosc. Radiat. Transf. **31**, 149–160 (1984)
54. Luikov, A.V., Bubnov, V.A., Soloviev, I.: On wave solutions of the heat-conduction equation. Int. J. Heat Mass Transf. **19**, 245–248 (1976)
55. Mongioví, M.S., Jou, D.: Thermodynamical derivation of a hydrodynamical model of inhomogeneous superfluid turbulence. Phys. Rev. B **75**, 024507 (14 pp.) (2007)
56. Mongioví, M.S., Jou, D., Sciacca, M.: Energy and temperature of superfluid turbulent vortex tangles. Phys. Rev. B **75**, 214514 (10 pp.) (2007)
57. Müller, I., Ruggeri, T.: Rational Extended Thermodynamics, 2nd edn. Springer, New York (1998)
58. Narayanamurti, V., Dynes, R.D.: Observation of second sound in bismuth. Phys. Rev. Lett. **28**, 1461–1465 (1972)
59. Peshkov, V.: Second sound in helium II. J. Phys. USSR **8**, 381–383 (1944)
60. Peshkov, V.: Determination of the velocity of propagation of the second sound in helium II. J. Phys. USSR **10**, 389–398 (1946)
61. Reissland, J.A.: The Physics of Phonons. Wiley, London (1973)
62. Saluto, L., Mongioví, M.S., Jou, D.: Longitudinal counterflow in turbulent liquid helium: velocity profile of the normal component. Z. Angew. Math. Phys. **65**, 531–548 (2014)
63. Sciacca, M., Sellitto, A., Jou, D.: Transition to ballistic regime for heat transport in helium II. Phys. Lett. A **378**, 2471–2477 (2014)
64. Sellitto, A., Cimmelli, V.A.: A continuum approach to thermomass theory. J. Heat Transf. - T. ASME **134**, 112402 (6 pp.) (2012)
65. Sellitto, A., Cimmelli, V.A.: Flux limiters in radial heat transport in silicon nanolayers. J. Heat Transf. - T. ASME **136**, 071301 (6 pp.) (2014)
66. Sellitto, A., Cimmelli, V.A., Jou, D.: Analysis of three nonlinear effects in a continuum approach to heat transport in nanosystems. Physica D **241**, 1344–1350 (2012)
67. Serdyukov, S.I.: Higher order heat and mass transfer equations and their justification in extended irreversible thermodynamics. Theor. Found. Chem. Eng. **47**, 89–103 (2013)
68. Sharma, K.R.: Damped Wave Transport and Relaxation. Elsevier, Amsterdam (2005)
69. Straughan, B.: Heat Waves. Springer, Berlin (2011)

70. Taitel, Y.: On the parabolic, hyperbolic and discrete formulation of the heat conduction equation. Int. J. Heat Mass Transf. **15**, 369–371 (1972)
71. Triani, V., Cimmelli, V.A.: Anisotropic heat transport in rigid solids. J. Non-Equilib. Thermodyn. **37**, 377–392 (2012)
72. Tzou, D.Y.: A unified field approach for heat conduction from micro-to-macro-scales. J. Heat Transf. - T. ASME **117**, 8–16 (1995)
73. Tzou, D.Y.: Nonlocal behavior in phonon transport. Int. J. Heat Mass Transf. **54**, 475–481 (2011)
74. Tzou, D.Y.: Longitudinal and transverse phonon transport in dielectric crystals. J. Heat Transf. - T. ASME **136**, 042401 (5 pp.) (2014)
75. Tzou, D.Y.: Macro- to Microscale Heat Transfer: The Lagging Behaviour, 2nd edn. Wiley, Chichester (2014)
76. Tzou, D.Y., Guo, Z.-Y.: Nonlocal behavior in thermal lagging. Int. J. Therm. Sci. **49**, 1133–1137 (2010)
77. Ván, P.: Weakly nonlocal irreversible thermodynamics. Ann. Phys. **12**, 146–173 (2003)
78. Ván, P., Fülöp, T.: Universality in heat conduction theory: weakly nonlocal thermodynamics. Ann. Phys. (Berlin) **524**, 470–478 (2012)
79. Van Sciver, S.W.: Helium Cryogenics, 2nd edn. Springer, Berlin (2012)
80. Vernotte, P.: Les paradoxes de la théorie continue de l'équation de la chaleur. C. R. Acad. Sc. **246**, 3154–3155 (1958)
81. Volz, S. (ed.): Thermal Nanosystems and Nanomaterials (Topics in Applied Physics). Springer, Berlin (2010)
82. Wang, M., Cao, B.-Y., Guo, Z.-Y.: General heat conduction equations based on the thermomass theory. Front. Heat Mass Transf. **1**, 013004 (8 pp.) (2010)
83. Wang, M., Yang, N., Guo, Z.-Y.: Non-Fourier heat conductions in nanomaterials. J. Appl. Phys. **110**, 064310 (7 pp.) (2011)
84. Ward, J.C., Wilks, J.: The velocity of second sound in liquid helium near the absolute zero. Philos. Mag. **42**, 314–316 (1951)
85. Zanchini, E.: Hyperbolic heat-conduction theories and nondecreasing entropy. Phys. Rev. B **60**, 991–997 (1999)
86. Zhang, Z.M.: Nano/Microscale Heat Transfer. McGraw-Hill, New York (2007)
87. Ziman, J.M.: Electrons and Phonons. Oxford University Press, Oxford (2001)

Chapter 3
Mesoscopic Description of Boundary Effects and Effective Thermal Conductivity in Nanosystems: Phonon Hydrodynamics

Nanometer-sized devices are of considerable current interest in micro/ nanoelectronics wherein the adjective "smaller" has meant greater performance ever since the invention of integrated circuits: more components per chips, faster operations, lower costs, and less power consumption. Miniaturization is also the trend in a range of other technologies, as for example, optics, catalysis, and ceramics. Many active efforts are currently observing in information storage, too, in order to develop magnetic and optical storage components with critical dimensions as small as tens of nanometers.

The raise of nanotechnology, indeed, requires increasing efforts to better understand the thermal-transport properties of nanodevices, as their performance and reliability are much influenced by memory, nonlocal and nonlinear effects [18, 44, 51, 57, 83, 87, 93]. Since the agreement between experiments and theory is still poor, the great challenge is to improve it. To reach this goal, the macroscopic derivation of generalized transport equations including memory, nonlocal and nonlinear effects represents a very important step. This problem may be tackled by different approaches, the most pursued of which is the microscopic one, based upon the Boltzmann equation.

Indeed, a very interesting approach is also that based on the so-called phonon hydrodynamics [21–23], which regards the whole set of heat carriers as a fluid whose hydrodynamic-like equations describe the heat transport [11]. This mesoscopic approach allows for a fast quantitative approximate estimation of the thermal properties, and may be a useful complement to microscopic theories, in order to select the most promising features of the nanosystems. In the linear regime the phonon hydrodynamics, wherein the phonons represent the main heat carriers, lays on the Guyer-Krumhansl (GK) transport equation (1.17) for the local heat flux $\mathbf{q}$ [1, 19, 20, 37, 38, 44, 82].

It is worth observing that although we often refer to Eq. (1.17) as the GK equation, it has in fact important differences with respect to the original proposal of Guyer and Krumhansl [1, 37, 38], especially regarding the way as the boundary

A. Sellitto et al., *Mesoscopic Theories of Heat Transport in Nanosystems*, SEMA SIMAI Springer Series 6, DOI 10.1007/978-3-319-27206-1_3

conditions are included in the model. In fact, those authors considered a boundary relaxation-time τ_b that they combined with the usual relaxation time due to bulk resistive mechanisms by means of Matthiessen rule as

$$\tau_R^{-1} = \tau_u^{-1} + \tau_i^{-1} + \tau_d^{-1} + \tau_b^{-1} \equiv \tau_{R0}^{-1} + \tau_b^{-1}$$

being τ_u the relaxation time of umklapp phonon-phonon collisions, τ_i the relaxation time of phonon-impurity collisions, and τ_d the relaxation time of phonon-defect collisions. Once the combined resistive-boundary (phonon-wall) collision time has been obtained, the thermal conductivity λ (depending on the size of the system through τ_b) was calculated, and used in the first term of the right-hand side of Eq. (1.17). When this is done, one obtains for the effective thermal conductivity λ_{eff} the expression (1.28), i.e.,

$$\lambda_{eff} = \frac{1}{3} c_v \bar{v}^2 \left(\frac{\tau_{R0}\tau_b}{\tau_{R0} + \tau_b} \right) = \lambda \left(1 + \frac{\tau_{R0}}{\tau_b} \right)^{-1} = \lambda \left[1 + f(T) \frac{\ell_p}{R} \right]^{-1} \tag{3.1}$$

wherein it has been considered $\ell_p = \tau_{R0}\bar{v}$, and $R = \tau_b\bar{v}/f(T)$, with $f(T)$ being a suitable temperature nondimensional function depending on the form of the cross section of the system, and whose order of magnitude is generally 1. This contribution adds, in some occasions, to the hydrodynamic contribution due to the normal (momentum-conserving) phonon-phonon collisions, which is described by the nonlocal term in the right-hand side of Eq. (1.17).

In our model, instead, we assume that thermal conductivity in Eq. (1.17) means the usual bulk thermal conductivity, and not the effective one as defined in Eq. (3.1). In doing this, we use an alternative approach which consists in including the boundary collision time τ_b not in the differential equation (1.17), but in suitable boundary constitutive equations [4, 5, 52, 74, 75]. To this end, it is important to note that in the phonon-hydrodynamic framework one has to pay more attention to boundary conditions [59], since the corresponding heat-transfer equation contains a nonlocal term of higher order than in the usual theories. Thus, in this new strategy we assume

$$\tau_R^{-1} \equiv \tau_{R0}^{-1} = \tau_u^{-1} + \tau_i^{-1} + \tau_d^{-1}$$

and focus our attention on the modelization of the constitutive equations for a slip heat flux [45] along the walls, necessary to complement Eq. (1.17). In such term, specular and diffusive phonon-wall collisions appear and, in rough walls, also some backscattering phonon-wall collisions.

From the practical point of view, the effective thermal conductivity in nanosystems (of different forms and shapes) is defined as

$$\lambda_{eff} \equiv \left(\frac{Q_{tot}}{A} \right) \frac{L}{\Delta T} \tag{3.2}$$

where L is the longitudinal length of the nanosystems, A is the area of its cross section, ΔT the temperature difference through the ends, and Q_{tot} the total heat per unit of time flowing in the system. Combining this effective thermal conductivity with the Fourier law, many practical problems of heat transport in nanosystems may be analyzed and, therefore, the analysis of such quantity plays a central role in nanotechnology. However, in the present book we will also see situations which are not truly describable by means of a Fourier-like expression with an effective thermal conductivity.

3.1 Phonon Slip Flow

In steady-state situations, complementing Eq. (1.17) with the local balance of the energy (1.8) (from which it follows $\dot{u} = 0 \Rightarrow \nabla \cdot \mathbf{q} = 0$), one obtains

$$\mathbf{q} = -\lambda \nabla T + \ell_p^2 \nabla^2 \mathbf{q} \tag{3.3}$$

as nonlocal constitutive equation for the heat flux. On the other hand, when the characteristic size of nanosystems is smaller than ℓ_p, the heat flux $\mathbf{q}$ can be neglected with respect to $\ell_p^2 \nabla^2 \mathbf{q}$ in Eq. (3.3), in such a way it reduces to [5, 7, 74, 75]

$$\nabla^2 \mathbf{q} = \frac{\lambda}{\ell_p^2} \nabla T. \tag{3.4}$$

Equation (3.4) is analogous to the Navier-Stokes equation for steady states, and for those situations in which the nonlinear convective term is negligible, i.e.,

$$\nabla^2 \mathbf{v} = \frac{1}{\eta} \nabla p \tag{3.5}$$

η being the shear viscosity of the fluid, $\mathbf{v}$ the velocity of the fluid, and p the pressure. The formal similarity between Eqs. (3.4) and (3.5) motivates the definition of "hydrodynamic regime" for those situations in which the heat flux obeys Eq. (3.4), and allows to identify the "viscosity" of phonons in terms of the thermal conductivity and of their mean-free path. Keeping in mind the very close forms of Eqs. (3.4) and (3.5), and since on microscopic ground the phonons may be viewed as a free-particle gas in a box [18, 84, 88], one may conclude that in the phonon-hydrodynamic framework $\mathbf{q}$, T and ℓ_p^2/λ play a role analogous to that played by $\mathbf{v}$, p, and η, respectively, in the fluid-dynamic framework [5, 7, 75]. Owing to this, in particular, it is also possible to refer to the ratio ℓ_p^2/λ as the "thermal viscosity". Note, however, that the concept of phonon hydrodynamics may also be given a more restrictive microscopic interpretation by relating it to collective, hydrodynamic-like, phonon behavior.

Since Eq. (3.4)—as well as Eq. (1.17)—is a second-order (in space) partial-differential equations, suitable boundary conditions are needed for its solution. We will assume for simplicity that the heat loss across the lateral walls of the systems is negligible. Nonetheless, due to the several phonon-wall collisions, a slip-flow contribution along the wall $\mathbf{q}_w$, additional to the bulk heat flux $\mathbf{q}_b$, should be expected [45, 53]. Both partial contributions to the overall local heat flux $\mathbf{q}$ are such that

$$\mathbf{q} = \mathbf{q}_b + \mathbf{q}_w. \tag{3.6}$$

In principle, the wall contribution $\mathbf{q}_w$ is restricted to a thin region near the walls, the so-called Knudsen layer, the thickness of which is of the order of mean-free path of the heat carriers. However, in a nanosystem whose characteristic dimension is comparable to (or smaller than) ℓ_p, the Knudsen layer pervades the whole system, and $\mathbf{q}_w$ may be assumed as a homogeneous contribution to the overall heat flux. The problem of heat slip flow along solid walls has been also investigated within the framework of modern thermodynamics in Refs. [45, 53] with the underlying idea of elevating the heat flux at the boundary to the status of independent variable, and in the phonon kinetic theory in Refs. [32, 91, 92].

To estimate the wall heat-flow contribution $\mathbf{q}_w$, helpful suggestions can be achieved from microfluidics [14]. This interdisciplinary field of research is concerned with the handling and transport of small amounts (nano/picoliters) of liquid. Microfluidic chips are already used in many laboratories for the considerable advantages derived from low-volume fluid treated [61]. In Table 3.1 the main analogies between integrated circuits and microfluidic chips are pointed out [8].

In microfluidics, the behavior of the fluid in the center of a flux-tube is ruled by the usual Navier-Stokes equations, whereas a slip flow is assumed along the surface. The two mostly-used types of boundary conditions for the slip flow read

$$\mathbf{v}_w = c'l\frac{\partial \mathbf{v}_b}{\partial \xi} \qquad \text{First-order slip-flow condition} \tag{3.7a}$$

$$\mathbf{v}_w = c'l\frac{\partial \mathbf{v}_b}{\partial \xi} - a'l^2\frac{\partial^2 \mathbf{v}_b}{\partial \xi^2} \quad \text{Second-order slip-flow condition} \tag{3.7b}$$

Table 3.1 Analogies between integrated circuits and microfluidic chips

	Integrated circuit	Microfluidic chip
Transported quantity	Energy	Mass
Material	Semiconductors (inorganic)	Polymers (organic)
Channel size	nm	μ m
Transport regime	Phonon hydrodynamics	Laminar fluid-dynamics

In the equations above $\mathbf{v}_b$ represents the fluid speed in the bulk of the microchannel, $\mathbf{v}_w$ is the velocity of the fluid on the walls, l is the mean-free path of the fluid particles, and ξ means the outward normal direction to the boundary. Moreover, c' and a' are positive constants, accounting for the features of the walls [16, 34, 78]. The first-order slip condition (3.7a) is the well-known Maxwell boundary condition [15, 47]. The second-order condition (3.7b), instead, was proposed in the decade of 1960 by several authors [16, 26], following the logics of the Knudsen gradient expansion to boundary conditions. It has recently received much attention in the domain of microfluidics of rarefied gases (see, for instance, Refs. [24, 39, 56, 62, 81, 90]). In particular, Eq. (3.7b) sets $a = 1/2$ in the so-called second-order slip model [41], and $a = 2/9$ in the so-called 1.5-model [63].

Due to the analogies phonon hydrodynamics/fluid dynamics, in Refs. [5, 7, 8, 46, 74–77] the following first- and second-order constitutive equations have been proposed for nanosystems

$$q_w = C\ell_p \left| \frac{\partial \mathbf{q}_b}{\partial \xi} \right| \tag{3.8a}$$

$$q_w = C\ell_p \left| \frac{\partial \mathbf{q}_b}{\partial \xi} \right| - \alpha \ell_p^2 \left| \frac{\partial^2 \mathbf{q}_b}{\partial \xi^2} \right| \tag{3.8b}$$

wherein C and α are positive constants, related to the properties of the walls. In more details, the coefficient C describes the specular and diffusive collisions of the phonons with the walls, whereas α accounts for phonon backscattering. Both coefficients are temperature dependent and account for the properties of the walls [74, 75], which may be smooth or rough [31].

These conditions may be used in a non-standard way, namely, assuming that the overall heat flux is given by Eq. (3.6) [5, 7, 46, 74–77]. In other words, in our approach we calculate the overall local heat flux $\mathbf{q}$ as follows: at first we obtain $\mathbf{q}_b$ by solving either Eq. (3.3), or Eq. (3.4) with vanishing heat flux on the boundary i.e., we assume that the bulk contribution to the local heat flux is the solution of generalized heat-transport equations. Then, we estimate the wall contribution $\mathbf{q}_w$ to the local heat flux by means of Eqs. (3.8). We explicitly note that in Eqs. (3.8) we introduce the absolute values for the spatial derivatives of $\mathbf{q}_b$, in contrast with the usual expressions in classical fluid dynamics for the boundary conditions, in order to emphasize that the first term (related to specular and diffusive reflections of heat carriers) therein has to turn out a contribution which goes in the same direction as $\mathbf{q}_b$, whereas the second term (related to backscattering, namely, to the backward reflections of heat carriers) have to turn out a contribution which goes in the opposite direction as $\mathbf{q}_b$.

Note that an alternative way to model the effective thermal conductivity would be to assume that the overall local heat flux, given by Eq. (3.6), is the solution of the generalized heat transport equation, and assuming that at the boundary of the system its value is such that

$$
q|_w = C\ell_p \left.\left|\frac{\partial \mathbf{q}}{\partial \xi}\right|\right|_w
$$
$$
q|_w = C\ell_p \left.\left|\frac{\partial \mathbf{q}}{\partial \xi}\right|\right|_w - \alpha\ell_p^2 \left.\left|\frac{\partial^2 \mathbf{q}}{\partial \xi^2}\right|\right|_w .
$$

Although this interpretation is not always equivalent to Eqs. (3.8), in most of the situations analyzed in the next it leads to equivalent results.

More general situations can be also treated following the way previously drawn to use Eqs. (3.8). For instance, in nanowires undergoing high-frequency perturbations, it is necessary to assume a relaxation of the heat flux on the boundary, too, through a dynamical constitutive equation which may be of the type [46]

$$
\tau_w \dot{q}_w + q_w = C\ell_p \left|\frac{\partial \mathbf{q}_b}{\partial \xi}\right| - \alpha\ell_p^2 \left|\frac{\partial^2 \mathbf{q}_b}{\partial \xi^2}\right| \tag{3.9}
$$

where τ_w represents the relaxation time accounting for the frequency of phonon-wall collisions, which here explicitly appears in contrast with Eqs. (3.8). Since these interactions may produce specular, diffusive and backward reflections of the phonons, the Matthiessen's rule would yields

$$
\tau_w^{-1} = \tau_{spec}^{-1} + \tau_{diff}^{-1} + \tau_{back}^{-1}
$$

where τ_{spec}, τ_{diff} and τ_{back} refer to the characteristic time of specular collisions, diffuse collisions and backscattering, respectively.

A qualitative estimation of τ_w may be obtained by evaluating the total frequency of collisions between phonons and walls. Indeed, since a wall, in principle, may show both smooth regions of width D, and rough regions of peaks Δ (see Fig. 3.1 for an illustrative sketch), in a first approximation it is possible to assume

$$
\frac{1}{\tau_w} = \frac{\overline{v}}{d}\left(\frac{D}{D+\Delta}\right) + \frac{\overline{v}}{d-\Delta}\left(\frac{\Delta}{D+\Delta}\right) \tag{3.10}
$$

wherein d is the characteristic size of the system, the ratio $D/(D+\Delta)$ indicates the probability of finding a smooth region, and $\Delta/(D+\Delta)$ the probability of finding a peak.[1] In the case of smooth walls (i.e., when $\Delta/D \to 0$), one has $\tau_w = d/\overline{v}$. Conversely, in the limit of very rough walls (i.e., when $\Delta \to d$), $\tau_w = 0$: in such a case phonons cannot advance in the nanowire because there is not enough free space to go ahead.

---

[1] We are assuming, for the sake of simplicity, that the width of the peaks is proportional to their height.

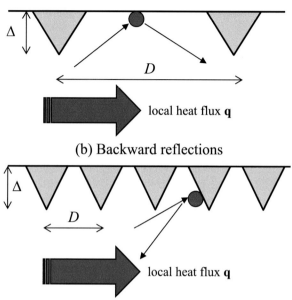

Fig. 3.1 Different degree of roughness of a wall, described by the parameters Δ (height of the roughness peaks) and D (separation of neighboring roughness peaks). The *red circle* sketches the phonon, and the *red arrow* indicates the direction of the local overall heat flux **q**. When the ratio Δ/D → 0 (or in the limit case when Δ = 0 nm), only specular and diffusive scattering are expected (**a**). Otherwise, backscattering is also expected (**b**). Furthermore, when D = Δ the surface has no flat regions, but it is completely rough

In the next sections we show the main results which can be obtained, in different situations, by applying to nanowires, nanotubes or thin layers, one of the non-standard constitutive equations for **q**$_w$ illustrated above.

3.2 Size Dependence of the Effective Thermal Conductivity

In this section, by means of our phonon-hydrodynamic approach, we will provide some theoretical expressions to model the effective thermal conductivity (3.2) in nanosystems of different geometries (i.e., circular cross section, core-shell and tubular nanowires, elliptical cross section, rectangular cross section and thin layers). For the sake of a didactic presentation, we start our analysis from the simplest cases which then will be generalized.

3.2.1 Nanowires with Circular Cross Sections

In order to show the way phonon hydrodynamics may allow to describe the size dependency of the effective thermal conductivity, we start by considering circular smooth-walled nanowires, with R as the radius of the (constant) transversal section, and L as the longitudinal length (see Fig. 3.2 for a qualitative sketch, as well as for the system of coordinates). These structures, which may be made of either metallic (e.g., Ni, Pt, Au), or semiconducting (e.g., Si, InP, GaN, etc.), or insulating (e.g., SiO_2, TiO_2) materials, in general have a thickness (or diameter) constrained to tens of nanometers, and are much used in current nanotechnology.

In steady-state situations and when $\mathbf{q} \ll \ell^2 \nabla^2 \mathbf{q}$, we previously observed that the GK equation (1.17) reduces to Eq. (3.4) which prescribes that the bulk heat-flow contribution has the following profile in each transversal section

$$q_b(r) = \frac{\lambda}{4\ell_p^2} \left(R^2 - r^2 \right) \frac{\Delta T}{L} \qquad (3.11)$$

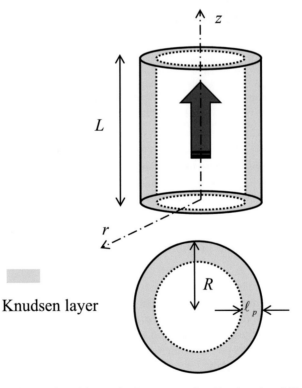

Fig. 3.2 Cylindrical nanowire with a R-circular cross section. For the sake of illustration, R is emphasized with respect to L. The heat (*red arrow*) is flowing along the longitudinal axis. In figure it is also shown the Knudsen layer near the walls. In this layer, the characteristic width of which is of the order of the phonon mean-free path, the wall contribution to the local heat flux is more relevant than outside

with r being the radial distance to the longitudinal axis of the nanowires, and the temperature gradient $\nabla T = -\Delta T/L$. In the case of smooth walls (i.e., in the absence of backscattering), the use of Eq. (3.8a) allows to obtain the following wall heat-flow contribution

$$q_w = C \left(\frac{\lambda R}{2\ell_p} \right) \frac{\Delta T}{L} \tag{3.12}$$

which, according to our approach summarized in Eq. (3.6), has to be added to Eq. (3.11). Note that Eq. (3.12) predicts a constant value for q_w in the whole transversal section. This is an approximation of our procedure, but which seems rather logical, since whenever Eq. (3.4) holds, the Knudsen layer pervades the whole transversal section [5, 74, 74].

Since the total heat per unit of time flowing in the system is

$$Q_{tot} = \int_0^R 2\pi r q\,(r)\,dr = \int_0^R 2\pi r\,(q_b + q_w)\,dr \tag{3.13}$$

according to Eq. (3.6), then the use of Eqs. (3.11) and (3.12) allows to obtain, from the definition (3.2), the following effective thermal conductivity [5]

$$\lambda_{eff}\,(Kn) = \frac{\lambda}{8\,Kn^2}\,(1 + 4C\,Kn) \tag{3.14}$$

wherein the Knudsen number $Kn = \ell_p/R$.

For high Kn values, the predicted theoretical effective thermal conductivity in Eq. (3.14) decreases linearly in terms of Kn^{-1}, according with experimental evidences [5]. It is easy to see that without the inclusion of the boundary heat flux in Eq. (3.8a) (i.e., when $C = 0$ in Eq. (3.14)), the effective thermal conductivity would decrease quadratically with respect to the reciprocal of the Knudsen number, against the experimental evidence.

Note, incidentally, that if one assumes, as in the original proposal of Guyer and Krumhansl [1, 37, 38], that in Eq. (1.17) λ is not the bulk thermal conductivity, but it is given as in Eq. (3.1), and no wall contribution to the local heat flux is taken into account (that is, $C = 0$), then Eq. (3.14) would lead to

$$\lambda_{eff}\,(Kn) = \frac{\lambda}{8\,(1 + a\,Kn)\,Kn^2}$$

which isn't able to describe experimental observations. Thus, that proposal must be interpreted in terms of the two-population models we drew in Sect. 2.4.4, which, in suitable conditions, would lead to

$$\lambda_{eff}\,(Kn) = \lambda \left[\frac{b\,(T)}{1 + a\,Kn} + \frac{1 - b\,(T)}{8\,Kn^2} \right]$$

with $b(T)$ being the relative population of resistive phonons, and $1 - b(T)$ the relative population of hydrodynamic ones. Furthermore, if one compares the high-Kn value of this expression with Eq. (3.14), then it will be possible to relate C with the ratio $b(T)/a(T)$. This shows how both alternative models may be compatible with each other. The advantage of our presentation is that the role of geometry is emphasized, whereas in the GK model (1.17) it is implicitly accounted in the coefficient $f(T)$.

The theoretical model (3.14) for λ_{eff} does no longer hold whenever the characteristic size of the nanowire (i.e., R) gets the same order of magnitude (or larger) of ℓ_p (i.e., if Kn ≤ 1). In fact, in this case, one has to use Eq. (3.3) in steady states, instead of Eq. (3.4).

Indeed, for vanishing values of the heat flux at the walls (i.e., $q_b(R) = 0$), the solution of Eq. (3.3) yields the following bulk heat flow profile [5, 77]

$$q_b(r) = \lambda \left[1 - \frac{J_0\left(ir/\ell_p\right)}{J_0\left(iR/\ell_p\right)} \right] \frac{\Delta T}{L} \tag{3.15}$$

with $J_0(z)$ being the zero-order cylindrical Bessel function of the indicated argument, whereas the nth-order cylindrical Bessel function is defined as

$$J_n(z) = \left(\frac{z}{2}\right)^n \sum_{t=0}^{\infty} \frac{(-1)^t \left(\frac{z}{2}\right)^{2t}}{t!\,(t+n)!}.$$

By imposing the first-order constitutive equation (3.8a), the wall contribution is

$$q_w(r) = -\lambda \left\{ C \left[\frac{iJ_1\left(iR/\ell_p\right)}{J_0^2\left(iR/\ell_p\right)} \right] J_0\left(ir/\ell_p\right) \right\} \frac{\Delta T}{L} \tag{3.16}$$

wherein $J_1(z)$ is the first-order cylindrical Bessel function. In deriving Eq. (3.16), we used the relation $J_0'(cz) = -cJ_1(cz)$, being c a generic constant. Note also that both $J_0\left(ir/\ell_p\right)$, and $iJ_1\left(iR/\ell_p\right)$ are functions defined on a pure imaginary field, but turning out only real values. Moreover, the minus in the right-hand side of Eq. (3.16) is due in order accounting for the positive value of q_w on the boundary. It is also worth noticing that in Eq. (3.16) q_w depends on the radial distance from the wall, since we are assuming now that R is comparable to ℓ_p.

Integrating the sum of Eqs. (3.15) and (3.16) across the transversal section, one finally gets the following effective thermal conductivity [77]

$$\lambda_{\text{eff}}(\text{Kn}) = \frac{\int_0^R 2\pi r \left[q_b(r) + q_w(r)\right] dr}{\pi R^2 (\Delta T/L)}$$

$$= \lambda \left\{ 1 - 2\,\text{Kn} \left[\frac{J_1\left(i/\,\text{Kn}\right)}{J_0\left(i/\,\text{Kn}\right)} \right]^2 \left[\frac{J_0\left(i/\,\text{Kn}\right)}{iJ_1\left(i/\,\text{Kn}\right)} + C \right] \right\}. \tag{3.17}$$

Straightforward calculations allow to see that whenever Kn gets high values, Eq. (3.17) reduces to Eq. (3.14).

3.2.2   Nanowires with Elliptical Cross Sections

Many studies have shown that the thermal conductivity of nanowires depends not only on the materials, but also on the cross-sectional size and shape [10, 42, 50, 77]. As a consequence of the fabrication processes, the cross sections of nanowires are generally elliptical, rather than circular [58].

Therefore, in the next we assume that the cross section of the nanowire is elliptical, and by means of phonon hydrodynamics we derive again the effective thermal conductivity in steady states. In particular, we assume $2b$ as the minor axis, and $2a$ as the major one (see Fig. 3.3). Moreover, the length L is supposed to be significantly larger than a, and the heat is flowing along the z-axis.

We assume that the nondimensional ratio between the characteristic size of the system and the mean-free path (i.e., the Knudsen number Kn) is bigger than unit. According with previously observations we may conclude that the heat carriers undergo the hydrodynamic regime whenever Kn $\gg$ 1. In practical applications, finding the characteristic size of a system is not so simple as it may appear. A wrong choice of this quantity may lead to a wrong model. Here we assume Kn $= \ell_p/b$ because the predominant phonon-wall collisions will be those corresponding to the thinner dimension. In this case, the solution of the hydrodynamic equation (3.4) for

Fig. 3.3 Nanowire with an elliptical cross section. For the sake of illustration, both sizes of the transversal section are emphasized with respect to the longitudinal length L. The system of coordinates is also shown, as well as the Knudsen layer pervading the whole transversal section

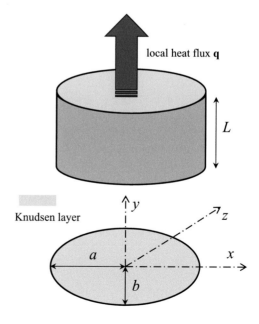

the bulk heat flux (i.e., for vanishing heat flux at the walls) is [8, 77]

$$q_b\,(x;y) = \frac{\lambda}{2\,Kn^2}\left(\frac{1}{1+\Phi^2}\right)\left(1-\frac{x^2}{a^2}-\frac{y^2}{b^2}\right)\frac{\Delta T}{L} \qquad (3.18)$$

where $\Phi = b/a$. Once Eq. (3.18) is introduced into Eq. (3.8a), it is simple to obtain that the wall contribution is

$$q_w\,(x) = \frac{\lambda}{Kn}\left(\frac{C}{1+\Phi^2}\right)\left[\sqrt{\frac{x^2}{a^2}(\Phi^2-1)+1}\right]\frac{\Delta T}{L}. \qquad (3.19)$$

Note that, although we analyze situations wherein $\ell_p \gg b$, here q_w cannot be considered as a homogeneous value across the transversal area: in the case of the elliptical cross section, in fact, the normal derivative does not have always the same value along the transversal wall.

The results in Eqs. (3.18) and (3.19) allow to derive, by direct calculations, the following expression for the effective thermal conductivity [77]:

$$\lambda_{eff}\,(Kn) = \frac{\displaystyle\int_{-b}^{b}\int_{-\sqrt{1-y^2/b^2}}^{\sqrt{1-y^2/b^2}}[q_b\,(x;y)+q_w\,(x)]\,dxdy}{\pi ab\,(\Delta T/L)}$$

$$= \frac{\lambda}{2\,Kn}\left[\frac{1}{2\,Kn}\left(\frac{1}{1+\Phi^2}\right)+CS\,(\Phi)\right] \qquad (3.20)$$

being $S\,(\Phi)$ the following numerical function [35, 77]

$$S\,(\Phi) = 1 - 0.6976\left(\frac{\Phi^2-1}{1.951\Phi^2+1}\right). \qquad (3.21)$$

Note that if $\Phi \equiv 1$ (that is, if the cross section is circular), Eq. (3.20) reduces to Eq. (3.20).

In Fig. 3.4 the theoretical behavior of the effective thermal conductivity in elliptical silicon nanowires at 150 K, arising from Eqs. (3.20)–(3.21), is plotted as a function of the ratio $\Phi = b/a$. The results in Fig. 3.4 show that the effective thermal conductivity is related to the shape of the cross section of the nanowires. In particular, the theoretical model in Eq. (3.20) prescribes that the smaller Φ, the greater λ_{eff}.

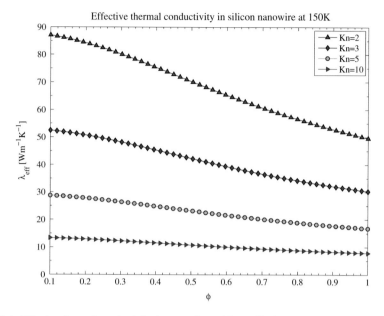

Fig. 3.4 Effective thermal conductivity in nanowires with an elliptical cross section as a function of Φ following from Eqs. (3.20)–(3.21). The nanowire is made of silicon at 150 K. Different values of the Knudsen number have been considered (that is, Kn = 2, Kn = 3, Kn = 5 and Kn = 10). Since in the theoretical proposal (3.20) we defined Kn = ℓ_p/b, changes in Φ are only due to changes in a

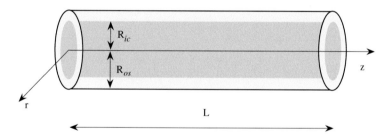

Fig. 3.5 Cylindrical concentric nanowire with a longitudinal heat flow. The *inner core* may be made either by different material with respect to the outer-shell (core-shell nanowire), or it may be vacuum (tubular nanowire)

3.2.3   Core-Shell and Tubular Smooth-Walled Nanowires

In this section we discuss the thermal conductivity of cylindrical concentric nanowires, such as core-shell nanowires or tubular nanowires (see Fig. 3.5), which are also used in several technological approaches [25].

In the limit of high Knudsen numbers, the heat flow profile for the inner core as a function of the radius, following from Eqs. (3.4) and (3.8a), is

$$q_{ic}(r) = \frac{\lambda_{ic}}{4\ell_{ic}^2} \frac{\Delta T}{L} \left(R_{ic}^2 - r^2 + 2C_{ic}\ell_{ic}R_{ic} \right) \tag{3.22}$$

wherein the subscript ic indicates that all the mentioned values are referred to the inner core. Since total heat flow along the inner nanowire is defined as

$$Q_{ic}^{(tot)} = \int_0^{R_{ic}} 2\pi r q_{ic}(r)\, dr$$

then direct calculations allow to obtain [7]

$$\lambda_{ic,\mathrm{eff}} = \frac{\lambda_{ic}}{8\,\mathrm{Kn}_{ic}^2} \left(1 + 4C_{ic}\,\mathrm{Kn}_{ic} \right) \tag{3.23}$$

for the inner-core effective thermal conductivity, being $\mathrm{Kn}_{ic} = \ell_{ic}/R_{ic}$ the corresponding Knudsen number, which corresponds to Eq. (3.14).

The local heat-flow profile for the outer shell (os), instead, has the form [49]

$$q_{os}(r) = -\frac{\lambda_{os}}{4\ell_{os}^2} \left(\frac{\Delta T}{L} \right) r^2 + A \ln r + B. \tag{3.24}$$

The parameters A and B may be found by using the boundary conditions as in Eq. (3.8a), i.e.,

$$q_{os}(R_{ic}) = C_{os}\ell_{os} \left(\frac{\partial q_{os}}{\partial r} \right)_{r=R_{ic}} \tag{3.25a}$$

$$q_{os}(R_{os}) = -C_{os}\ell_{os} \left(\frac{\partial q_{os}}{\partial r} \right)_{r=R_{os}}. \tag{3.25b}$$

The difference of signs in the equations above is due to the fact that at $r = R_{ic}$, $q_{os}(r)$ is an increasing function of r, whereas at $r = R_{os}$ it is a decreasing function of r. For this reason we prefer to write explicitly both conditions in a separate way. Observe that the coefficients C_{ic} and C_{os} in principle could be different. This means that the surfaces at the two walls in contact with the outer shell are assumed to have the same material features (namely, the same coefficient C_{os}), and the inner surfaces of the separation wall are assumed to have the coefficient C_{ic}.

The solution of Eqs. (3.24)–(3.25) leads to [7]

$$A = - \frac{\lambda_{os}}{16\,\mathrm{Kn}_{os}^2} \left(\frac{\Delta T}{L} \right) \Gamma'$$
(3.26a)

$$B = \frac{\lambda_{os}}{32\,\mathrm{Kn}_{os}^2} \left(\frac{\Delta T}{L} \right) \left[\frac{1 + \Phi_{cs}^2}{(1 - \Phi_{cs})^2} + 4C_{os}\,\mathrm{Kn}_{os} \right] - \frac{A}{2} \left(\ln R_{os} + \ln R_{ic} \right)$$

$$+ \frac{A}{\Phi_{cs}^2} \left[C_{os}\Phi_{cs} (1 - \Phi_{cs})^2\,\mathrm{Kn}_{os} \right]$$
(3.26b)

$$\Gamma' = \frac{\Phi_{cs}^2 (1 + \Phi_{cs})(1 + 4C_{os}\,\mathrm{Kn}_{os})}{(1 - \Phi_{cs}) \left[\Phi_{cs} (1 - \Phi_{cs}^2)\, 2C_{os}\,\mathrm{Kn}_{os} + (1 - \Phi_{cs})^2\,\mathrm{Kn}_{os}^2 - \Phi_{cs}^2 \ln \Phi_{cs} \right]}$$
(3.26c)

where $\Phi_{cs} = R_{ic}/R_{os}$, and $\mathrm{Kn}_{os} = \ell_{os}/ [2\,(R_{os} - R_{ic})]$ is the Knudsen number for the outer shell.

Since total heat flow along the outer nanowire is defined as

$$Q_{os}^{(tot)} = \int_{R_{ic}}^{R_{os}} 2\pi r q_{os}(r)\, dr$$

from Eqs. (3.24) and (3.26) the following effective thermal conductivity for the outer shell arises [7]:

$$\lambda_{os,\mathrm{eff}} = \frac{\lambda_{os}}{16\,\mathrm{Kn}_{os}^2} \left\{ -\frac{\Gamma'}{2} \left[1 + \left(\frac{1 + \Phi_{cs}^2}{1 - \Phi_{cs}^2} \right) \ln \Phi_{cs} \right] \right.$$

$$\left. + 2C_{os}\,\mathrm{Kn}_{os} \left[1 + \frac{\Gamma'(1 - \Phi_{cs})^2}{2} \frac{}{\Phi_{cs}} \right] \right\}$$
(3.27)

which also represents the effective thermal conductivity ($\lambda_{t,\mathrm{eff}}$) of a tubular nanowire with R_{ic} as the inner radius, and R_{os} as the outer radius, whereas the effective thermal conductivity of the core-shell nanowire may be obtained by Eqs. (3.23) and (3.27) as

$$\lambda_{nw,\mathrm{eff}} = \lambda_{ic,\mathrm{eff}}\Phi_{cs}^2 + \lambda_{os,\mathrm{eff}} \left(1 - \Phi_{cs}^2 \right).$$
(3.28)

3.2.4 Thin Layers and Nanowires with Rectangular Cross Sections

The phonon-hydrodynamic approach also allows to derive a theoretical model for the effective thermal conductivity in thin layers, or in nanosystems the cross section of which shows a rectangular shape (see Fig. 3.6).

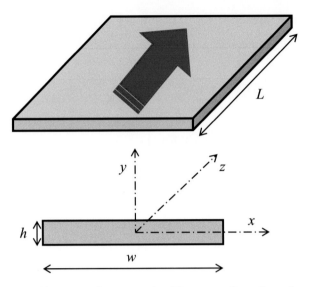

Fig. 3.6 Thin layer with a rectangular cross section. The system of coordinates in each transversal section is such that $x \in [-w/2; w/2]$, and $y \in [-h/2; h/2]$. The length L of the layer is much larger than h and w. The heat is flowing along the longitudinal axis z

For small values of Kn (defined in this case as $\mathrm{Kn} = \ell_p/h$), if the width w of the cross section is much larger than its thickness h, then Eq. (3.3) yields the following bulk heat-flow profile (i.e., when the heat flux at the walls vanishes) [5]

$$q_{\mathrm{b}}\,(y) = \lambda \left[1 - \frac{\cosh\left(\frac{y}{\ell_p}\right)}{\cosh\left(\frac{1}{2\,\mathrm{Kn}}\right)} \right] \frac{\Delta T}{L} \qquad (3.29)$$

where y is the axis perpendicular to the parallel plates, and spans the range from $-h/2$ to $h/2$. The use of the boundary condition (3.8a) turns out that the wall contribution in this case is

$$q_{\mathrm{w}}\,(y) = \lambda \left[C \tanh\left(\frac{1}{2\,\mathrm{Kn}}\right) \frac{\cosh\left(\frac{y}{\ell_p}\right)}{\cosh\left(\frac{1}{2\,\mathrm{Kn}}\right)} \right] \frac{\Delta T}{L} \qquad (3.30)$$

and, consequently, the effective thermal conductivity reads [77]

$$\lambda_{\mathrm{eff}}\,(\mathrm{Kn}) = \lambda \left\{ 1 - 2\,\mathrm{Kn}\tanh\left(\frac{1}{2\,\mathrm{Kn}}\right) \left[1 - C \tanh\left(\frac{1}{2\,\mathrm{Kn}}\right) \right] \right\}. \qquad (3.31)$$

For increasing values of Kn, Eq. (3.31) tends to the limit value [5]

$$\lambda_{\text{eff}} (\text{Kn}) = \frac{\lambda}{12 \, \text{Kn}^2} (1 + 6C \, \text{Kn}) \tag{3.32}$$

which predicts a linear decrease of the effective thermal conductivity in terms of Kn^{-1} whenever $\text{Kn} \to \infty$, still complying with experimental measurements [9, 13, 54, 55, 80].

If one relaxes the hypothesis $w \gg h$, the following expression for the effective heat conductivity can be achieved [77]

$$\lambda_{\text{eff}} (\text{Kn}) = \frac{\lambda}{2 \, \text{Kn}} \left\{ \frac{1}{6 \, \text{Kn}} \left[1 - \frac{192}{\pi^5} \gamma \sum_{t,\text{odd}}^{\infty} \frac{1}{t^5} \tanh \left(\frac{t\pi}{2\gamma} \right) \right] \right.$$
$$\left. + C \left[1 - \frac{8}{\pi^3} \gamma \sum_{t,\text{odd}}^{\infty} \frac{1}{t^3} \tanh \left(\frac{t\pi}{2\gamma} \right) \right] \right\} \tag{3.33}$$

γ being the nondimensional ratio h/w. Up to the first-order approximation in γ, Eq. (3.33) reduces to

$$\lambda_{\text{eff}} (\text{Kn}) = \frac{\lambda}{12 \, \text{Kn}^2} \left[1 - 0.630\gamma + 6C \, \text{Kn} \, (1 - 0.271\gamma) \right] . \tag{3.34}$$

Note the very close behavior of the theoretical prediction of λ_{eff} in Eq. (3.32) and in Eq. (3.34), the sole difference between these behaviors resting on the presence of the shape factor γ. Increasing values of γ yield a reduction in the effective thermal conductivity, for a given value of h (or of Kn), because this means a reduction of w and, therefore, an increase of the phonon-wall collisions with the lateral walls.

Whenever $\gamma \equiv 1$, Eq. (3.34) turns out the effective thermal conductivity in the case of square cross section, i.e.,

$$\lambda_{\text{eff}} (\text{Kn}) = \frac{\lambda}{32.4 \, \text{Kn}^2} (1 + 12C \, \text{Kn}) . \tag{3.35}$$

At the very end, let us observe that in thin layers the effective thermal conductivity along the layer (i.e., that for an in-plane heat transfer) is not the same as that perpendicular to the layer (i.e., that for a cross-plane heat transfer) which also includes the thermal resistance of the walls [28].

3.2.5 Thin Channels Filled with Superfluid Helium

As well as thermal waves, phonon hydrodynamics was first discovered in superfluid liquid helium [12, 29, 43, 69]. Here we will only give a brief introduction to the problem of heat transport along thin tubes, which is in fact one of the richest and most challenging topics in heat transfer [29, 85, 89].

In steady states, the heat transfer in turbulent liquid helium [64, 65, 71–73] may be described by an equation analogous to Eq. (3.3), namely, by

$$\alpha L k \mathbf{q} = -A \nabla T + B \nabla^2 \mathbf{q} \qquad (3.36)$$

where $A = \rho_n \rho T S^2$ with ρ_n as the mass density of the normal component of the fluid and ρ the total mass density, and $B = \eta_n$ as the viscosity of the normal component. In the framework of EIT, superfluid helium is considered as a single fluid with the heat flux as an internal variable [64]. In the more usual two-fluid model by Landau and Tisza [36, 48, 49, 73] it is considered to be composed of a normal viscous component and a superfluid one. Their densities, respectively, are ρ_n and ρ_s in such a way that $\rho = \rho_n + \rho_s$. The velocities of two components, instead, are $\mathbf{v}_n$ and $\mathbf{v}_s$, and are related to the barycentric velocity $\mathbf{v}$ as $\mathbf{v} = \rho_n \mathbf{v}_n + \rho_s \mathbf{v}_s$.

In the coefficient A in Eq. (3.36) S means the specific entropy which is carried out by the normal component, and the heat flux is identified as $\mathbf{q} = \rho_n T S \mathbf{v}_n$. Thus, the two terms in the right-hand side of Eq. (3.36), in steady states, come from the hydrodynamic equation $\nabla p = \eta_n \nabla^2 \mathbf{v}_n$ for the viscous component, complemented by the thermodynamic relation $S dT = v dp$, from which it follows $\nabla p = \rho S \nabla T$.

The term of the left-hand side of Eq. (3.36), instead, accounts for the internal friction between normal fluid and the quantized vortices arising in the turbulent. Therein, L is the vortex length density, k is the quantum of vorticity (given by $k = h/m_{He}$ with h being the Planck constant, and m_{He} the atomic mass of helium), and α is a friction coefficient. The value of L is zero for small velocities, namely, when $v_n d/k < \mathrm{Re}_{q1}$ with d as the diameter of the channel and Re_{q1} a critical value of the quantum Reynolds number.

Usually, one takes for $\mathbf{v}_n$ (and therefore, for $\mathbf{q}$) the non-slip boundary conditions. Nevertheless, at sufficiently low temperatures and sufficiently narrow channels, when the phonon mean-free path becomes comparable to the diameter of the channel, a slip flow arises [73]. However, the problem is more complicated than in solid, because L itself is proportional to the square of the heat-flux modulus for sufficiently large values of $\mathbf{q}$, the situation in which the turbulence is fully developed. For a detailed treatment on this topic we refer the readers to [71, 72].

3.3 Phonon Backscattering and Roughness Dependence of the Effective Thermal Conductivity

In previous section we observed that boundary conditions may be interpreted in microscopic terms on the basis of phonon-wall interactions, which is an active topic of research in heat transport in nanosystems [13, 53, 80]. The relevance of such boundary conditions has been fostered by observations of a drastic reduction in the thermal conductivity in rough-walled nanowires as compared with that of smooth-walled ones [9, 40, 60]. In this way, the analysis of boundary conditions becomes a crucial aspect of heat transport in nanosystems, and a challenge for thermodynamic descriptions [2, 3, 86]. The role of boundary conditions is also an interesting problem in extended thermodynamics [30, 44, 51, 53, 67, 70, 78], which has higher-order moments of the distribution function as independent variables of the description, in addition to the classical thermodynamic variables.

To extend to rough-walled nanosystems (wherein the phenomenon of backscattering may occur), the theoretical proposal in Eq. (3.8b) can be used to evaluate the wall contribution to the local heat flux. From the physical point of view, the negative sign in the second term (derived from kinetic theory) allows the possibility to describe phonon backscattering at the walls, because it gives a contribution which has an opposite sign to that of the first term, which goes in the same direction as the bulk heat flux. For some specific geometries of the rugosity of the wall, the heat carriers (i.e., the phonons, in this case) could bounce backwards, and go in a direction opposite to that of the bulk heat flux (see Fig. 3.1).

In the case of a nanowire with a circular cross section, in steady states and whenever the radius is much smaller than the phonon mean-free path (i.e., $\mathrm{Kn} > 1$), the combination of the wall condition (3.8b) and the parabolic heat profile (3.11) gets

$$q_{\mathrm{w}} = \frac{\lambda}{2\ell_p} \left(CR - \alpha\ell_p \right) \frac{\Delta T}{L} \qquad (3.37)$$

which becomes negative for values of the mean-free path such that $\ell_p > \ell_c$, being $\ell_c = CR/\alpha$ a critical mean-free path (or, alternatively, for radius R smaller than $\alpha\ell_p/C$).

Consequently, the local overall heat-flow profile $q(r)$ is given by

$$q(r) = \frac{\lambda}{4\ell_p^2} \left(R^2 - r^2 + 2C\ell_p R - 2\alpha\ell_p^2 \right) \frac{\Delta T}{L} \qquad (3.38)$$

which predicts the possibility of finding a negative value of the local heat flux in some regions of the system.

Referring the readers to Sect. 3.3.3 for a deeper analysis of this situation, let us now focus our attention on effective thermal conductivity, which from Eq. (3.38), turns out to be [6]

$$\lambda_{\mathrm{eff}} = \frac{\lambda}{8\,\mathrm{Kn}^2} \left(1 + 4C\,\mathrm{Kn} - 4\alpha\,\mathrm{Kn}^2 \right) \qquad (3.39)$$

wherein backscattering is related to the coefficient α, whereas specular and diffusive scattering are related to the coefficient C (see Fig. 3.1).

3.3.1 Roughness Dependence of the Wall Coefficients

Since C and α play a very important role, it is natural to go deeper in their physical insight. To do this, we start to assume that the roughness of the wall is described by two parameters: Δ which is the root-mean square value (rms) of the roughness fluctuations, and D which is the average distance between roughness peaks [31]. In terms of these parameters the coefficients C and α can be modeled as [75]

$$\begin{cases} C = C' \left(1 - \dfrac{\Delta}{D}\right) \\ \alpha = \alpha' \dfrac{\Delta}{D} \end{cases} \tag{3.40}$$

where C' and α' are numerical dimensionless functions, dependent on temperature, but independent on Δ and D. Indeed, when $\Delta/D \to 0$ (or in the limit case when $\Delta = 0$ nm), no backscattering is expected, but only specular and diffusive scattering. Therefore $\alpha \to 0$ when $\Delta/D \to 0$. On the other side, when $D = \Delta$, the surface has no flat regions, but it is completely rough, in which case $C = 0$ is expected. In Eqs. (3.40) we have assumed the simplest dependence for C and α on Δ/D. Another interesting proposal for that coefficient may be found in Ref. [90]. Moreover, in Eqs. (3.40) the coefficients C' and α' depend, instead, on temperature, because a surface is considered smooth (or rough) when the characteristic height Δ of the roughness is smaller (or higher) than the dominant phonon wavelength, which depends on temperature [75]. This dependence, as well as the dependence on temperature of λ and ℓ_p, is necessary to describe the temperature dependence of the effective thermal conductivity.

When the assumptions (3.40) are made, the theoretical prediction in Eq. (3.39) becomes

$$\lambda_{\text{eff}} = \frac{\lambda}{8\,\text{Kn}^2}\left(1 + 4C'\,\text{Kn}\right) - \frac{\lambda}{2\,\text{Kn}}\left(\frac{\Delta}{D}\right)\left(C' + \alpha'\,\text{Kn}\right). \tag{3.41}$$

In order to face the theoretical proposal in Eq. (3.41) with experimental observations, we use the effective thermal conductivity for smooth and rough Si nanowires of different (circular) cross section [40, 54, 60]. Some values of λ_{eff} at different temperatures, both in the absence of backscattering and in its presence, arising from experimental data in silicon nanowires are quoted in Tables 3.2 and 3.3, respectively. In particular, in Table 3.2 three different radii are considered (i.e., $R = 115$ nm, $R = 56$ nm, and $R = 37$ nm), whereas in Table 3.3 two other different radii are considered (i.e., $R = 115$ nm, and $R = 97$ nm). We refer to Table 1.1 for the experimental values of the bulk thermal conductivity and of phonon mean-free path of silicon at several temperatures. In that table, ℓ_p has been obtained from λ by using the relation $\lambda = c_v \bar{v} \ell_p / 3$. The obtained value of ℓ_p will be a characteristic mean-free path incorporating the several bulk collisions, but not the

Table 3.2 Experimental values of the effective thermal conductivity (W/mK) in silicon nanowires with different radii R (nm) and at several values of temperature (K) in the absence of backscattering ($\Delta = 0$ nm)

	$T = 150$	$T = 100$	$T = 80$	$T = 60$	$T = 50$	$T = 40$	$T = 30$
R	λ_{eff}	λ_{eff}	λ_{eff}	λ_{eff}	λ_{eff}	λ_{eff}	λ_{eff}
115	46	45	40	27	19	13	5
56	28	23	21	16	11	7	3
37	17	14	11	8	6	4	1.7

The experimental data have been inferred from Ref. [54]

Table 3.3 Experimental values of the effective thermal conductivity (W/mK) in silicon nanowires with different radii R (nm) and at several values of temperature (K) in the presence of backscattering ($\Delta = 3$ nm, and $D = 6$ nm)

	$T = 150$	$T = 100$	$T = 80$	$T = 60$	$T = 50$	$T = 40$	$T = 30$
R	λ_{eff}	λ_{eff}	λ_{eff}	λ_{eff}	λ_{eff}	λ_{eff}	λ_{eff}
115	7.8	5.7	4.9	3.3	2.5	1.8	1.3
97	5.3	3.8	3.2	2.1	1.7	1.2	0.9

The experimental data have been inferred from Refs. [40, 60]

boundary collisions. Remember that, from a microscopic point of view, the phonon mean-free path depends on the phonon frequency and on the kind of collisions one is considering. The characteristic value used here is the simplest estimation from macroscopic experimental data.

As a tentative model of the variations of C' and α' with temperature in the interval between 30 and 150 K we take the following functions

$$C'(T) = C_3 T^3 + C_2 T^2 + C_1 T + C_0 \tag{3.42a}$$

$$\alpha'(T) = \alpha_3 T^3 + \alpha_2 T^2 + \alpha_1 T + \alpha_0 \tag{3.42b}$$

wherein an empirical fit with experimental observation suggests

$$\begin{cases} C_3 = 4.0 \cdot 10^{-6}\,\text{K}^{-3}, \ C_2 = -1.1 \cdot 10^{-3}\,\text{K}^{-2}, \ C_1 = 9.4 \cdot 10^{-2}\,\text{K}^{-1}, \quad C_0 - 1.5 \\ \alpha_3 = 1.1 \cdot 10^{-7}\,\text{K}^{-3}, \ \alpha_2 = -6.4 \cdot 10^{-6}\,\text{K}^{-2}, \ \alpha_1 = 4.6 \cdot 10^{-4}\,\text{K}^{-1}, \ \alpha_0 = -1.0 \cdot 10^{-2}. \end{cases} \tag{3.43}$$

The use of Eq. (3.41), joined with the values of C' and α' obtained by Eqs. (3.42) and (3.43), allows to predict the results of Table 3.4 in the absence of backscattering, and those of Table 3.5 in the presence of backscattering. The results reasonably fit with experimental data.

The value of the effective thermal conductivity, arising from the model above, in the case of a nanowire with $R = 115$ nm at 150 K is very different from the experimental one, both in the absence of backscattering, and in the presence of it. This is a logical consequence of the simplifying assumptions we made to derive Eq. (3.41). That discrepancy, in fact, is due to the relative small value of

Table 3.4 Theoretical predictions for the value of the effective thermal conductivity (W/mK) in silicon nanowires with different radii R (nm) and at several values of temperature (K) in the absence of backscattering ($\Delta = 0$ nm)

	$T = 150$	$T = 100$	$T = 80$	$T = 60$	$T = 50$	$T = 40$	$T = 30$
R	λ_{eff}	λ_{eff}	λ_{eff}	λ_{eff}	λ_{eff}	λ_{eff}	λ_{eff}
115	68.3	45.5	40.3	27.6	19.0	11.7	6.4
56	28.1	20.9	19.4	13.4	9.2	5.7	3.1
37	17.5	13.6	12.7	8.8	6.1	3.8	2.1

These values are obtained by using Eq. (3.41)

Table 3.5 Theoretical predictions for the value of the effective thermal conductivity (W/mK) in silicon nanowires with different radii R (nm) and at several values of temperature (K) in the presence of backscattering ($\Delta = 3$ nm, and $D = 6$ nm)

	$T = 150$	$T = 100$	$T = 80$	$T = 60$	$T = 50$	$T = 40$	$T = 30$
R	λ_{eff}	λ_{eff}	λ_{eff}	λ_{eff}	λ_{eff}	λ_{eff}	λ_{eff}
115	15.1	6.9	6.5	4.2	2.3	1.3	2.0
97	5.4	2.4	3.1	2.0	0.8	0.4	1.5

These values are obtained by using Eq. (3.41)

the corresponding Knudsen number, which in this case is Kn $= 1.57$. Remember that Eq. (3.41) is only valid for Kn $\gg 1$, and when Kn ≤ 1 Eq. (3.17) should be used instead of Eq. (3.14). Since the mean-free path becomes larger at lower temperatures, instead, the Knudsen number becomes sufficiently high for lower temperatures even for the nanowire of radius 115 nm and, of course, for all the thinner nanowires. This justifies the good agreement between the experimental data and the theoretical ones in all the other cases.

3.3.2 Effective Phonon Mean-Free Path and Nonlocal Effects

Note that in Eq. (3.3) one could have alternatively written

$$\mathbf{q} = -\lambda \nabla T + \tilde{a}^2 \ell_p^2 \nabla^2 \mathbf{q} \tag{3.44}$$

with $\tilde{a}(T)$ being a dimensionless function of temperature, and ℓ_p still indicating the characteristic phonon mean-free path following from the relation $\lambda = (1/3) c_v \bar{v} \ell_p$. In this case, instead of Eq. (3.8a) for example, one would have

$$q_{\text{w}} = \tilde{a} C \ell_p \left| \frac{\partial \mathbf{q}_{\text{b}}}{\partial \xi} \right| \tag{3.45}$$

and two adjustable functions to fit experimental data (that is, $\widetilde{a}(T)$ and $C(T)$) may be used. In previous section we have chosen to use the simplest proposal, namely, $\widetilde{a}(T) \equiv 1$ with $C(T)$ (and also $\alpha(T)$ in the case of rough walls with backscattering) with the sole adjustable parameter. This is sufficient for a plausible description of experimental results, as we previously underlined. Indeed, for an higher precision, one may also use the other adjustable parameter $\widetilde{a}(T)$. In this case, one can take for the parameter C in Eq. (3.8) the microscopic expression [5, 14, 17, 79, 82, 91])

$$C = 2\left(\frac{1+P}{1-P}\right) \tag{3.46}$$

with P (which may vary between 0 and 1) being the relative contribution of specular phonon-wall collisions as compared with the total number of specular and diffusive phonon-wall collisions. The dependence of P with temperature is relatively small. Since P is not known a priori, but depends on the manufacturing technique, C should also be determined from experimental data, but, if the interpretation (3.46) is used for C, it should be higher than 2, and the dimensionless function $\tilde{a}(T)$ should be different than 1. If this is done, one finds values of $\tilde{a}(T)$ between 5 and 15 in the case of silicon, thus indicating that the choice in Eq. (3.44)—alternative to Eq. (3.3)—could be closer to some proposals in which the characteristic length related to nonlocal effects is one order of magnitude higher than the effective mean-free path derived from the bulk thermal conductivity.

Note further that the consequent *re-scaled* value of the phonon mean-free path $\widetilde{a}(T)\, \ell_p(T)$ would be a consequence of the fact that phonons of different frequencies contribute in a different form to the first and second term of Eq. (3.3). In other words, in a more microscopic setting one could consider Eq. (3.3) in term of a phonon frequency ω_p as

$$\mathbf{q}(\omega_p) = -\lambda(\omega_p)\nabla T + \ell_p^2(\omega_p)\nabla^2 \mathbf{q}(\omega_p)$$

$$= -\frac{c_v(\omega_p)\,\overline{v}(\omega_p)\,\ell_p(\omega_p)}{3}\nabla T + \ell_p^2(\omega_p)\nabla^2 \mathbf{q}(\omega_p) \tag{3.47}$$

with $\mathbf{q}(\omega_p)$ being the contribution of phonons of frequency ω_p to the local heat flux, in such a way that $\mathbf{q} = \int \mathbf{q}(\omega_p)\,n(\omega_p)\,d\omega_p$, and $\lambda = \int \lambda(\omega_p)\,n(\omega_p)\,d\omega_p$, where $n(\omega_p)$ is the number of phonons of frequency between ω_p and $\omega_p + d\omega_p$ per unit volume of the system. When integrating Eq. (3.47) over the frequency, one would obtain Eq. (3.45) with $\tilde{a} \neq 1$.

Here we will not deal longer with this aspect, which should deserve more attention in future analyses.

3.3.3 Phonon Backscattering and Conductor-Insulator Transition

The use of the boundary condition (3.8b) pointed out that the theoretical heat-flow profile may be not everywhere positive, as one would naively imagine. In fact, from Eq. (3.38) one recovers that there may be some points wherein the local overall heat flow results to be negative. In particular, for mean-free path values larger than the critical value ℓ_c, the local heat flux will be negative in the annular region defined by

$$R\sqrt{1 - \frac{2\alpha\,(\ell_c + \Delta\ell)\,\Delta\ell}{R^2}} < r < R \qquad (3.48)$$

with $\Delta\ell = \ell_p - \ell_c$. From the physical point of view, in this annular region the heat flux goes from lower temperature to the higher one.

In Fig. 3.7 the overall local heat-flow profile in nanowires is qualitatively plotted as a function of the radius r. The parabolic profile of Fig. 3.7a is still obtained by assuming a zero tangential heat flux on the walls. In Fig. 3.7b, instead, the effects of the boundary condition (3.8b) have been taken into account, and the overall local heat flow does not longer assume positive values everywhere, i.e., there is an annular region wherein it goes backwards.

It seems useful to note that a negative value of $q\,(r)$ in the annular region (3.48) does not imply that the total heat flow Q_{tot} along the system is negative. In the present case, from Eq. (3.38), it results in

$$Q_{\text{tot}} = \int_0^R 2q\,(r)\,dr = \frac{\pi R^4 \lambda}{8\ell_p^2}\left[1 + 2C\frac{\ell_p}{R} - 2\alpha\left(\frac{\ell_p}{R}\right)^2\right]\frac{\Delta T}{L} \qquad (3.49)$$

the corresponding effective thermal conductivity being given by Eq. (3.39). For Kn lower than the critical value

$$\text{Kn}_c = \frac{C}{2\alpha}\left(\sqrt{1 + \frac{\alpha}{C}} + 1\right) \qquad (3.50)$$

the effective thermal conductivity (3.39) is positive, and vanishes for $\text{Kn} \geq \text{Kn}_c$.

It is interesting to compare now the results of the first-order slip model ($\alpha = 0$) with those of the higher-order slip model ($\alpha \neq 0$). Indeed, if the ratio R/ℓ_p becomes smaller than $1/\text{Kn}_c$ the effective thermal conductivity (3.39) vanishes, as a negative value is not physically consistent with the global formalism of the second law. This implies that if the radius R of the nanowire is such that the relation

$$R < R_c = \frac{2\alpha\ell_p}{C}\left(\sqrt{1 + \frac{\alpha}{C}} + 1\right)^{-1} \qquad (3.51)$$

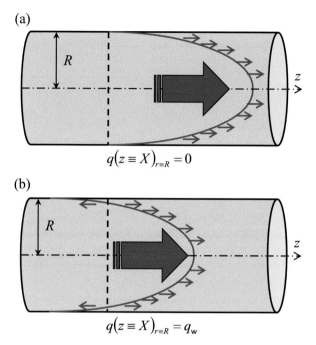

Fig. 3.7 Local heat-flow profile in nanowires. The longitudinal heat flow vanishes on the walls (**a**). The *vertical dash-dotted line X* in both figures represents the transversal section wherein the local heat flux is evaluated. In the presence of a non-vanishing heat flow $\mathbf{q_w}$ on the lateral walls, the heat-flow profile does not start from $\mathbf{q} = 0$ on the walls (**b**), but from $\mathbf{q_w}$. Since in this case the local heat-flow profile may start from a negative value (due to phonon backscattering), there is a region near the walls wherein it results to be negative, namely the heat locally flows from the coldest region to the hottest one

holds, then the system at hand is no longer a heat conductor, but becomes an insulator. This recalls the idea of the metal-insulator Anderson transition [27, 33]. This transition is found in amorphous systems, where the atomic disorder is so high that extended electron states become suppressed, and only localized electron states remain, which do not allow a long-range electron current. In our case, and on microscopic grounds, this feature would be found when the backwards reflections of the phonons on the walls (which are related to the coefficient α as observed above) are so important that heat cannot progress along the nanowire, i.e., only localized phonon states are possible. A deeper consideration of this analogy, at the microscopic level, could be interesting.

Another physical possibility of reducing the heat flux is the localization due to quantum effects. Indeed, when the radius of the nanowire becomes very small, in view of the Heisenberg principle it follows that the transversal speed of the particles must become high, thus reducing the energy going along the nanowire, as a fraction of the total energy will be removed from the longitudinal dimension to feed the energy of the transversal direction. Thus, even in smooth walls, for sufficiently thin

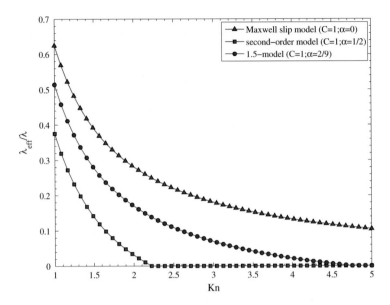

Fig. 3.8 Comparison of the behavior of the effective thermal conductivity (3.39) in a silicon nanowire in terms of the Knudsen number, for three different slip boundary conditions. The figure plots the results predicted by the Maxwell slip model [i.e., Eq. (3.8a)] and those predicted by higher-order slip models [i.e., second-order slip model (3.8b) with $\alpha = 1/2$, and $\alpha = 2/9$]

nanowires, quantum localization effects could strongly reduce the total heat flux beyond the reduction implied by classical collisions of phonons with the walls.

In Fig. 3.8 we compare the effective thermal conductivity following from the first-order condition (3.8a) with that from second-order conditions (3.8b). For the sake of illustration, in that figure we have supposed $C = 1$, while $\alpha = 0$, $\alpha = 1/2$ and $\alpha = 2/9$. Moreover, we supposed that the nanowire is made of silicon at 300 K. In Fig. 3.8 the conductor/insulator transition can be clearly seen whenever Eq. (3.8b) is used.

It is also interesting to note that for Knudsen numbers such that

$$\frac{C}{2\alpha} \leq \mathrm{Kn} \leq \mathrm{Kn}_c \tag{3.52}$$

there will be a local zone wherein the classical thermodynamic restriction for the entropy production

$$\sigma_{\mathrm{le}}^{(s)} = \mathbf{q} \cdot \nabla T^{-1} \geq 0 \tag{3.53}$$

is not fulfilled, even if λ_{eff} is positive. In other words, the second law is fulfilled in the bulk, but it is apparently violated in the external annular region (3.48). That way, the arising question is whether this situation is admissible, or not. Indeed, this problem is circumvented if one looks at the theories beyond local-equilibrium, incorporating nonlocal terms, which show a non-trivial contribution to the entropy production. In particular, in EIT [44, 51] the entropy production is

$$\sigma_{\text{eit}}^{(s)} = \mathbf{q} \cdot \nabla T^{-1} + \frac{\ell_p^2}{\lambda T^2} \mathbf{q} \cdot \nabla^2 \mathbf{q} + \frac{\ell_p^2}{\lambda T^2} \nabla \mathbf{q} : \nabla \mathbf{q} \qquad (3.54)$$

namely, it is locally allowed that the Eq. (3.53) be not fulfilled, without violation of second law of thermodynamics. In particular, introducing the parabolic heat profile (3.38) in Eq. (3.54) we have

$$\sigma_{\text{eit}}^{(s)} = \frac{\lambda}{4\ell_p^2} \left(\frac{\Delta T}{T} \right)^2 \left(\frac{r}{L} \right)^2 \qquad (3.55)$$

that is, the entropy production is everywhere positive in the system at hand. Thus, the boundary condition (3.8b) is admissible in the framework of EIT.

3.3.4   A Qualitative Microscopic Interpretation of C' and α'

The forms (3.42) for the coefficients C' and α' are useful from a practical point of view, but they are not illustrative about the microscopic processes leading to a modification of such parameters with the temperature T. Thus, in this section, we will try to understand the most relevant aspects of this dependence. To do so, first of all, it is convenient to plot the behavior of both coefficients in terms of temperature, as described by Eqs. (3.42) and (3.43). In Fig. 3.9 the behaviors of $C'(T)$ and $\alpha'(T)$ are shown. The most salient features in Fig. 3.9 are the presence of a maximum for $C'(T)$ around 60 K, and the increase of $\alpha'(T)$ beyond a temperature of the order of 30 K. Note that the two sketches in Fig. 3.9 have a different vertical scale, the one for $C'(T)$ going from 0.1 to 0.9, and the one for $\alpha'(T)$ going from 0 to 0.35. Now, we will try to qualitatively understand these two outstanding features.

A relevant quantity in the phonon-roughness interaction is the phonon wavelength. The most probable phonon wavelength λ_{phonon} is given by a value of the order of $\lambda_{\text{phonon}} \approx 8.5 \cdot 10^{-8}$ mK $/T$ (K). This expression comes from the form of the Planck distribution function for phonons, and is the analogous of Wien law for the most probable photon wavelength λ_{photon}, namely, $\lambda_{\text{photon}} \approx 2.9 \cdot 10^{-3}$ mK $/T$(K), but with the sound velocity in the material (which for silicon is of the order of $8.5 \cdot 10^3$ m/s), instead of the speed of light in vacuo.

The effects of phonon backscattering are expected to be especially relevant when the most probable phonon wavelength λ_{phonon} is lower than the root-mean square of the roughness fluctuations Δ. Since λ_{phonon} decreases for increasing temperature, it is expected that the backscattering term should increase for increasing temperature. For low temperatures, such that $\lambda_{\text{phonon}} > \Delta$, the wall should appear smooth to phonons, and α' should vanish, practically. Since in the experimental values we are referring to $\Delta = 3$ nm, the temperature T_0 below which the roughness should

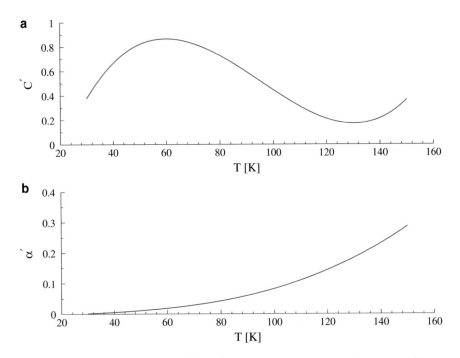

Fig. 3.9 Behavior of the coefficients C' and α' in Eqs. (3.40) as functions of temperature (figures a and b, respectively). These results, arising from Eqs. (3.42) and (3.43), refer to nanowires made by silicon

become irrelevant will be of the order of the ratio $8.5 \cdot 10^{-8}\,\mathrm{mK}/3 \cdot 10^{-9}\,\mathrm{m}$, namely, $T_0 \approx 27\,\mathrm{K}$. This value is of the order of the temperature value below which Fig. 3.9b yields very small values for α'. It is expected that after a deep increase of this parameter for rising temperatures, the value of α' will reach an asymptotic value when $\lambda_{\mathrm{phonon}}$ is sufficiently smaller than Δ (let us say, an order of magnitude smaller, implying a temperature of the order of 270 K). The temperature dependence of $\alpha'(T)$ could alternatively be assumed to be $\alpha'(T) = \alpha'_0$ for $T < T_0$, with $T_0 = 8.5 \cdot 10^{-8}\,\mathrm{mK}/\Delta\,(\mathrm{m})$, and $\alpha'(T) = \alpha'_0 + \alpha'_1\,(T - T_0)^2$, up to a saturation value for temperatures higher than $10T_0$. Note that we take this indicative value because it corresponds to a characteristic phonon wavelength of the order $\lambda_{\mathrm{phonon}} \approx \Delta/10$, an order of magnitude less than the peak rugosity. We comment about this simple guess only to illustrate how the form of $\alpha'(T)$ could depend on Δ.

The maximum in the coefficient $C'(T)$ could be tentatively interpreted as a resonance between the typical phonon wavelength and the typical separation length D between rugosity peaks of the wall. The peak in Fig. 3.9a is found approximatively at $T = 60\,\mathrm{K}$, whereas the vanishing of $\alpha'(T)$ in Fig. 3.9b is approximatively found at $T = 30\,\mathrm{K}$. Remembering that the experimental values for λ_{eff} in Table 3.3 correspond to $D = 6\,\mathrm{nm}$, we have that the mentioned resonance would be expected for $\lambda_{phonon} \approx 2L \approx 12\,nm$. However, using the above expression for λ_{phonon}, one obtains a temperature of the order of 7 K, i.e., one order of magnitude smaller than the observed peak at 60 K. This discrepancy could be partially mitigated if one takes into account that D is an average separation distance between peaks, whose actual separation distance may be considerably smaller than this value. If the resonance is related not to the average value between higher peaks, but to the actual separation between peaks, the resonance would appear at higher temperature and would be relatively wide. Thus, whereas the temperature at which the coefficient $\alpha'(T)$ begins to increase is connected to the height of the roughness peaks, the maximum in the coefficient $C'(T)$ would be related to the minimum separation between peaks. In this sense, the coefficients $C'(T)$ and $\alpha'(T)$ are not expected to be universal, but dependent on the features of the roughness and the phonon wavelength.

According to the previous interpretation, it could seem that C' should not depend on temperature for smooth walls, because the resonance effects mentioned above would be lost. However, there is another source of temperature variation of the coefficient C'. The diffuse scattering should depend on temperature, as it is the result of the re-emission of the particle from the wall with a probability distribution function for the velocity which corresponds to the equilibrium distribution function with the temperature of the wall. In contrast, the specular collisions are not expected to depend on temperature. This makes that the relative role of specular and diffusive collisions changes with temperature. This dependence could be studied by analyzing the variation of the coefficient $C'(T)$ with temperature for smooth walls. After having this information, one should superpose to it the possibility of a resonance due to the effects discussed above.

The behavior of $C'(T)$ in Fig. 3.9a arises from Eq. (3.42a) with the identification in Eqs. (3.43), i.e., by taking into account the data of smooth- as well as of rough-walled nanowires. The result is satisfactory from the point of view of the conductivity results. However, one could ask what would be the difference for $C'(T)$ if it is computed separately for smooth-walled (i.e., when $\Delta = 0\,nm$, or when $D \to \infty$) and for rough-walled nanowires (i.e., when $\Delta \neq 0\,nm$), because, according to our interpretation, the geometry of the roughness has a relevant role in the temperature dependence of $C'(T)$. Thus, we have obtained again $C'(T)$ separately for smooth- and rough-walled nanowires. In the former case, instead of Eq. (3.42a), we have supposed

$$C'(T) = C_4 T^4 + C_3 T^3 + C_2 T^2 + C_1 T + C_0 \qquad (3.56)$$

where the comparison with the experimental data in Table 3.2 allows us to obtain that $C_4 = -2.9 \cdot 10^{-8}\,K^{-4}$, $C_3 = 1.4 \cdot 10^{-5}\,K^{-3}$, $C_2 = -2.2 \cdot 10^{-3}\,K^{-2}$, $C_1 = 1.4 \cdot 10^{-1}\,K^{-1}$, and $C_0 = -2.3$. With this type of parametrization, it is possible to obtain the values of the effective thermal conductivity of Table 3.6. These values still well fit with the experimental data in Table 3.2. However, these results are only slightly different from those shown in Table 3.4, and probably indistinguishable from them once the experimental error bars are taken into account. Thus, the parametrization in Eq. (3.56) has mainly a theoretical interest, in showing

Table 3.6 Effective thermal conductivity (W/mK) in silicon nanowires at several values of temperature and for three different radii (nm) for smooth-walled nanowires (i.e., when $\Delta = 0$ nm, or when $D \to \infty$)

R	$T = 150$ λ_{eff}	$T = 100$ λ_{eff}	$T = 80$ λ_{eff}	$T = 60$ λ_{eff}	$T = 50$ λ_{eff}	$T = 40$ λ_{eff}	$T = 30$ λ_{eff}
115	68.0	47.4	38.6	27.4	19.4	12.1	6.1
56	27.9	21.9	18.5	13.3	9.4	6.0	3.0
37	17.4	14.2	12.1	8.7	6.2	3.9	1.9

The values are obtained by using Eqs. (3.41) and (3.56), namely, the coefficient C' is not influenced by the rugosity

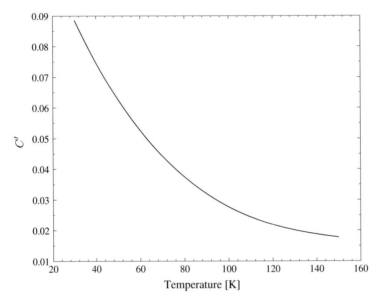

Fig. 3.10 Behavior of the coefficient C' as function of temperature in the case of smooth walls (i.e., when $\Delta = 0$ nm, or when $D \to \infty$). This behavior follows from Eq. (3.56). Since in this case the separation length between roughness peaks tends to infinite, the resonance manifests its effects only at very low temperatures

that the qualitative behavior of $C'(T)$ for smooth-walled nanowires agrees with the expected features following from our simple qualitative interpretation of the resonance between phonons and roughness-peak separations. On the practical side, the parametrization in Eq. (3.42a) shown in Fig. 3.9 and interpreted in this section, may be satisfactorily used.

In Fig. 3.10 we plot the behavior of $C'(T)$ arising from Eq. (3.56). The behavior of $C'(T)$ plotted in Fig. 3.10 is consistent with our interpretation of a resonance related to the minimum separation between roughness peaks. In smooth-walled nanowires this separation tends to infinite, and therefore this resonance will take place for extremely low temperatures, tending to zero. Thus, it is logical to have a maximum at very low temperature, and decreasing values for higher temperature. Furthermore, this behavior is also consistent with the Ziman formula for the specularity parameter p for a surface with random roughness of height Δ, namely, $p = \exp\left[-16\pi^3\left(\Delta/\lambda_{\text{phonon}}\right)^2\right]$, assuming that Δ is not strictly zero, but simply very small, and which predicts a fast decay of $C'(T)$ for increasing temperature [66, 94]. Indeed, it is logical to expect that $C'(T)$ should be approximately proportional to the specularity parameter, which is zero for totally diffuse scattering. In the latter case, the number of forwards and backwards phonons emitted by the wall is equal, implying a vanishing heat flux along the wall. Since at high temperatures diffuse scattering predominates over the specular one, the decrease of $C'(T)$ is to be expected. However, in the Ziman model there is no backscattering, and this implies vanishing wall heat flux. When one considers backscattering effects described by the coefficient α given by Eq. (3.40), instead, one may have a phonon back-flow on the walls. Of course, the maximum value of this back-flow cannot be higher than the forwards phonon flow driven by the longitudinal temperature gradient.

3.4 Phonon-Wall Interactions and Frequency-Dependent Thermal Conductivity in Nanowires

Memory effects may drastically influence the behavior of nanodevices at high frequencies [2, 68, 76]. In the simplest description of relaxational effects in the bulk for phonon heat transport, the heat flux $\mathbf{q}$ is given by the MCV equation (1.6), the Fourier transform of which leads to an effective frequency-dependent thermal conductivity of the form

$$\lambda_{\text{eff}}(\omega) = \frac{\lambda}{1 + (\omega\tau_R)^2}. \qquad (3.57)$$

Equation (3.57) points out that the frequency-dependent thermal conductivity is reduced with respect to that of steady states. In fact, since the steady-state thermal conductivity is of the order of $\lambda = (1/3)\,c_v\bar{v}^2\tau_R$, an increase in τ_R yields both an increase of it (which is linear in the relaxation time τ_R and independent of frequency ω), but it also increases the denominator of Eq. (3.57) (in a quadratic way for high enough frequency) and therefore it reduces the thermal conductivity. In nanowires, the situation is still more complex and interesting. We already observed that the steady-state thermal conductivity is much reduced with respect to the bulk thermal conductivity. In the high-frequency regime, there is a further reduction analogous to that appearing in Eq. (3.57), but more complicated, because it will also depend on the collisions with the lateral walls, which are not included in Eq. (3.57). Thus, to have a detailed estimation of high-frequency thermal conductivity in nanosystems it is needed to go beyond the usual steady-state analyses and incorporate both the inertia of heat, and the dynamical aspects of the phonon collisions with the walls.

The frequency-dependent effective thermal conductivity was calculated in Ref. [3] for cylindrical nanowires by using a continued-fraction expansion for the frequency and wavevector-dependent thermal conductivity. However, in Ref. [3] the role of the lateral surfaces was not considered.

Phonon hydrodynamics also allows a more intuitive description both of phonon transport, and of the role of phonon-wall collisions in frequency-dependent situations. To show this, let us consider the following sinusoidal varying perturbations for the difference in temperatures (applied at the ends of a nanowire) and for the bulk heat flow:

$$\Delta T (\omega; t) = \Delta \tilde{T} e^{i\omega t} \tag{3.58a}$$

$$q_{\mathrm{b}} (r; \omega; t) = \tilde{q}_{\mathrm{b},0} \left(\frac{R^2 - r^2}{4\ell_p^2} \right) e^{i\omega t} \tag{3.58b}$$

The form in Eq. (3.58a) considers that the temperature only depends on the longitudinal position, but it is homogeneous across every transversal area. The form in Eq. (3.58b) for the perturbation of the flux, instead, assumes that the bulk heat-flow profile keeps essentially the parabolic form corresponding to the Poiseuille phonon flow, but with an amplitude changing periodically, instead of being directly given by the steady value $-\lambda \nabla T$, as in the steady-state solution. Note that the parabolic form for the heat flux profile in nonequilibrium state is not necessary, but it is a simplifying assumption. In fact, the detailed form of the profile is not practically very relevant, because in realistic situations only the total heat flux Q_{tot} across the whole transversal section may be measurable, but not its radial dependence. Indeed, one could also assume that the bulk heat flow depends on space and expand this variation in spatial Fourier series across the cylinder. This would contribute with higher harmonics to the decay, which could be studied in the future, but which presumably will have less relevance than the lowest-order term we are studying here.

When Eqs. (3.58) are used in the combination of the evolution equation of the internal energy per unit volume (1.8) with the GK equation (1.17), then the following unsteady bulk heat-flow profile arises [76]:

$$q_{\mathrm{b}} (r; \omega; t) = \left(\lambda + 2i\omega c_v \ell_p^2 \right) \left[\frac{R^2 - r^2}{4\ell_p^2 + i\omega \tau_R (R^2 - r^2)} \right] \frac{\Delta \tilde{T}}{L} e^{i\omega t}. \tag{3.59}$$

In view of the dynamical form of the bulk heat-flow contribution (3.59), in the same way let us suppose that the wall contribution is given by

$$q_{\mathrm{w}} (\omega; t) = \tilde{q}_{\mathrm{w},0} e^{i\omega t} \tag{3.60}$$

wherein $\tilde{q}_{\mathrm{w},0}$ can be obtained from the boundary condition in Eq. (3.9), with the relaxation time τ_{w} therein given by Eq. (3.10) [76]. In this way, from Eq. (3.2) one has

$$
\begin{aligned}
\lambda_{\mathrm{eff}}(\omega; \mathrm{Kn}) &= \mathrm{Re}\left\{\frac{Q_{\mathrm{tot}}}{\pi R^2}\frac{L}{\Delta T}\right\} \\
&= \frac{2\,\mathrm{Kn}^2}{\omega^2\tau_R^2}\left[\lambda\ln\left(1+\frac{\omega^2\tau_R^2}{16\,\mathrm{Kn}^4}\right)-4\omega c_v\ell_p^2\arctan\left(\frac{\omega\tau_R}{4\,\mathrm{Kn}^2}\right)\right] \\
&\quad + \frac{2c_v\ell_p^2}{\tau_R}\frac{C}{2\,\mathrm{Kn}}\left(\frac{\lambda+2\omega^2\tau_{\mathrm{w}}c_v\ell_p^2}{1+\omega^2\tau_{\mathrm{w}}^2}\right) \\
&\quad - \frac{\alpha}{2\left(1+\omega^2\tau_{\mathrm{w}}^2\right)}\left[\lambda\left(1+\frac{\omega^2\tau_R\tau_{\mathrm{w}}}{\mathrm{Kn}^2}\right)+2\omega^2 c_v\ell_p^2\left(\tau_{\mathrm{w}}-\frac{\tau_R}{\mathrm{Kn}^2}\right)\right].
\end{aligned}
$$

$$\tag{3.61}$$

It is easy to observe that in the low-frequency limit (i.e., when $\omega\tau_R \to 0$), the frequency-dependent effective thermal conductivity above reduces to the value of the effective thermal conductivity in steady states (3.39), whereas in the high-frequency limit (i.e., when $\omega\tau_R \to \infty$) one has

$$
\lambda_{\mathrm{eff}}(\mathrm{Kn}) = \frac{2c_v\ell_p^2}{\tau_R}\left\{1+\frac{1}{2}\frac{\tau_R}{\tau_{\mathrm{w}}}\left[\frac{C}{\mathrm{Kn}}-\alpha\left(1-\frac{\tau_R}{\tau_{\mathrm{w}}}\frac{1}{\mathrm{Kn}^2}\right)\right]\right\}-\frac{\alpha}{2}\frac{\lambda}{\mathrm{Kn}^2}\frac{\tau_R}{\tau_{\mathrm{w}}}. \tag{3.62}
$$

In any way, as the role of τ_{w} is usually not considered in the frequency-dependent thermal conductivity, one should be cautious with these preliminary results, until the dynamical aspects of phonon collisions with the walls are better understood. Indeed, from Eq. (3.61) it is possible to observe that the effect of τ_{w} is stronger than that of τ_R itself. This is logical, because in the situation we are studying the collisions of the phonons against the walls are much more frequent than those with other phonons or of impurities in the bulk.

References

1. Ackerman, C.C., Guyer, R.A.: Temperature pulses in dielectric solids. Ann. Phys. **50**, 128–185 (1968)
2. Alvarez, F.X., Jou, D.: Memory and nonlocal effects in heat transports: from diffusive to ballistic regime. Appl. Phys. Lett. **90**, 083109 (3 pp.) (2007)
3. Alvarez, F.X., Jou, D.: Size and frequency dependence of effective thermal conductivity in nanosystems. J. Appl. Phys. **103**, 094321 (8 pp.) (2008)
4. Alvarez, F.X., Jou, D.: Boundary conditions and evolution of ballistic heat transport. J. Heat Transf. T. ASME **132**, 012404 (6 pp.) (2009)
5. Alvarez, F.X., Jou, D., Sellitto, A.: Phonon hydrodynamics and phonon-boundary scattering in nanosystems. J. Appl. Phys. **105**, 014317 (5 pp.) (2009)

6. Alvarez, F.X., Jou, D., Sellitto, A.: Pore-size dependence of the thermal conductivity of porous silicon: a phonon hydrodynamic approach. Appl. Phys. Lett. **97**, 033103 (3 pp.) (2010)
7. Alvarez, F.X., Jou, D., Sellitto, A.: Phonon boundary effects and thermal conductivity of rough concentric nanowires. J. Heat Transf. T. ASME **133**, 022402 (7 pp.) (2011)
8. Alvarez, F.X., Cimmelli, V.A., Jou, D., Sellitto, A.: Mesoscopic description of boundary effects in nanoscale heat transport. Nanoscale Systems MMTA **1**, 112–142 (2012)
9. Asheghi, M., Leung, Y.K., Wong, S.S., Goodson, K.E.: Phonon-boundary scattering in thin silicon layers. Appl. Phys. Lett. **71**, 1798–1800 (1997)
10. Barnard, A.S., Curtiss, L.A.: Modeling the preferred shape, orientation and aspect ratio of gold nanorods. J. Mater. Chem. **17**, 3315–3323 (2007)
11. Bausch, W.: Thermal conductivity and Poiseuille flow of phonons in dielectric films and plates. Phys. Status Solidi B **52**, 253–262 (1972)
12. Benin, D., Maris, H.J.: Phonon heat transport and Knudsen's minimum in liquid helium at low temperatures. Phys. Rev. B **18**, 3112–3125 (1978)
13. Bergmann, G.: Conductance of a perfect thin film with diffuse surface scattering. Phys. Rev. Lett. **94**, 106801 (3 pp.) (2005)
14. Bruus, H.: Theoretical Microfluidics. Oxford University Press, Oxford (2007)
15. Burgdorfer, A.: The influence of the molecular mean path on the performance of hydrodynamic gas lubricated bearings. J. Basic Eng. T. ASME **81**, 94–100 (1959)
16. Cercignani, C.: Higher order slip according to the linearized Boltzmann equation. California University Berkeley Institute of Engineering Research, Report AS-64-19, Berkeley (1964)
17. Cercignani, C.: Rarefied Gas Dynamics. Cambridge University Press, Cambridge (2000)
18. Chen, G.: Nanoscale Energy Transport and Conversion - A Parallel Treatment of Electrons, Molecules, Phonons, and Photons. Oxford University Press, Oxford (2005)
19. Cimmelli, V.A., Frischmuth, K.: Nonlinear effects in thermal wave propagation near zero absolute temperature. Phys. B **355**, 147–157 (2005)
20. Cimmelli, V.A., Frischmuth, K.: Gradient generalization to the extended thermodynamic approach and diffusive-hyperbolic heat conduction. Phys. B **400**, 257–265 (2007)
21. Cimmelli, V.A., Sellitto, A., Jou, D.: Nonlocal effects and second sound in a nonequilibrium steady state. Phys. Rev. B **79**, 014303 (13 pp.) (2009)
22. Cimmelli, V.A., Sellitto, A., Jou, D.: Nonequilibrium temperatures, heat waves, and nonlinear heat transport equations. Phys. Rev. B **81**, 054301 (9 pp.) (2010)
23. Cimmelli, V.A., Sellitto, A., Jou, D.: Nonlinear evolution and stability of the heat flow in nanosystems: beyond linear phonon hydrodynamics. Phys. Rev. B **82**, 184302 (9 pp.) (2010)
24. Colin, S., Lalonde, P., Caen, R.: Validation of a second-order slip flow model in rectangular microchannels. Heat Transf. Eng. **25**, 23–30 (2004)
25. Cui, L.-F., Yang, Y., Hsu, C.-M., Cui, Y.: Carbon-Silicon Core-shell nanowires as high capacity electrode for lithium ion batteries. Nano Lett. **9**, 3370–3374 (2009)
26. Deissler, R.G.: An analysis of second-order slip flow and temperature-jump boundary conditions for rarefied gases. Int. J. Heat Mass Transf. **7**, 681–694 (1964)
27. Dobrosavljević, V., Kotliar, G.: Mean field theory of the Mott-Anderson transition. Phys. Rev. Lett. **78**, 3943–3946 (1997)
28. Dong, Y., Cao, B.-Y., Guo, Z.-Y.: Ballistic-diffusive phonon transport and size induced anisotropy of thermal conductivity of silicon nanofilms. Phys. E **66**, 1–6 (2015)
29. Donnelly, R.J.: Quantized Vortices in Helium II. Cambridge University Press, Cambridge, UK (1991)
30. Dreyer, W., Struchtrup, H.: Heat pulse experiments revisited. Contin. Mech. Thermodyn. **5**, 3–50 (1993)
31. Ferry, D.K., Goodnick, S.M.: Transport in Nanostructures, 2nd edn. Cambridge University Press, Cambridge (2009)
32. Fryer, M.J., Struchtrup, H.: Moment model and boundary conditions for energy transport in the phonon gas. Cont. Mech. Thermodyn. **26**, 593–618 (2014)
33. García-García, A.M.: Classical intermittency and quantum Anderson transition. Phys. Rev. E **69**, 066216 (2004)

34. Gombosi, T.I.: Gaskinetic Theory. Cambridge University Press, Cambridge (1994)
35. Graur, I., Sharipov, F.: Gas flow through an elliptical tube over the whole range of gas rarefaction. Eur. J. Mech. B/Fluids **27**, 335–345 (2008)
36. Greywall, D.S.: Thermal-conductivity measurements in liquid ^{4}He below 0.7k. Phys. Rev. B **23**, 2152–2168 (1981)
37. Guyer, R.A., Krumhansl, J.A.: Solution of the linearized phonon Boltzmann equation. Phys. Rev. **148**, 766–778 (1966)
38. Guyer, R.A., Krumhansl, J.A.: Thermal conductivity, second sound and phonon hydrodynamic phenomena in nonmetallic crystals. Phys. Rev. **148**, 778–788 (1966)
39. Hadjiconstantinou, N.G.: Comment on Cercignani's second-order slip coefficient. Phys. Fluids **15**, 2352 (3 pp.) (2003)
40. Hochbaum, A.I., Chen, R., Delgado, R.D., Liang, W., Garnett, E.C., Najarian, M., Majumdar, A., Yang, P.: Enhanced thermoelectric performance of rough silicon nanowires. Nature **451**, 163–167 (2008)
41. Hsia, Y.T., Domoto, G.A.: An experimental investigation of molecular rarefaction effects in gas lubricated bearings at ultra-low clearance. J. Lubr. Technol T. ASME **105**, 120–130 (1983)
42. Hui, L., Wang, B.L., Wang, J.L., Wang, G.H.: Local atomic structures of palladium nanowire. J. Chem. Phys. **121**, 8990–8996 (2004)
43. Jou, D., Lebon, G., Mongioví, M.S.: Second sound, superfluid turbulence, and intermittent effects in liquid helium II. Phys. Rev. B **66**, 224509 (9 pp.) (2002)
44. Jou, D., Casas-Vázquez, J., Lebon, G.: Extended Irreversible Thermodynamics, 4th revised edn. Springer, Berlin (2010)
45. Jou, D., Lebon, G., Criado-Sancho, M.: Variational principles for thermal transport in nanosystems with heat slip flow. Phys. Rev. E **82**, 031128 (6 pp.) (2010)
46. Jou, D., Sellitto, A., Alvarez, F.X.: Heat waves and phonon-wall collisions in nanowires. Proc. R. Soc. A **467**, 2520–2533 (2011)
47. Kennard, E.H.: Kinetic Theory of Gases. McGraw-Hill, New York (1938)
48. Landau, L.D.: The theory of superfluidity of He II. J. Phys. **60**, 356–358 (1941)
49. Landau, L.D., Lishitz, E.M.: Mechanics of Fluids. Pergamon, Oxford (1985)
50. Law, M., Goldberger, J., Yang, P.: Semiconductor nanowires and nanotubes. Annu. Rev. Mater. Res. **34**, 83–122 (2004)
51. Lebon, G., Jou, D., Casas-Vázquez, J.: Understanding Nonequilibrium Thermodynamics. Springer, Berlin (2008)
52. Lebon, G., Machrafi, H., Grmela, M., Dubois, C.: An extended thermodynamic model of transient heat conduction at sub-continuum scales. Proc. R. Soc. A **467**, 3241–3256 (2011)
53. Lebon, G., Jou, D., Dauby, P.C.: Beyond the Fourier heat conduction law and the thermal non-slip condition. Phys. Lett. A **376**, 2842–2846 (2012)
54. Li, D., Wu, Y., Kim, P., Shi, L., Yang, P., Majumdar, A.: Thermal conductivity of individual silicon nanowires. Appl. Phys. Lett. **83**, 2934–2936 (2003)
55. Liu, W., Asheghi, M.: Phonon-boundary scattering in ultrathin single-crystal silicon layers. Appl. Phys. Lett. **84**, 3819–3821 (2004)
56. Lockerby, D.A., Reese, J.M., Emerson, D.R., Barber, R.W.: Velocity boundary condition at solid walls in rarefied gas calculations. Phys. Rev. E **70**, 017303 (2004)
57. Luzzi, R., Vasconcellos, A.R., Galvão Ramos, J.: Predictive Statistical Mechanics: A Nonequilibrium Ensemble Formalism. Fundamental Theories of Physics. Kluwer, Dordrecht (2002)
58. Majumdar, A., Lin, C.-H.: Gate capacitance of cylindrical nanowires with elliptical cross-sections. Appl. Phys. Lett. **98**, 073506 (3 pp.) (2011)
59. Márkus, F., Gambár, K.: Heat propagation dynamics in thin silicon layers. Int. J. Heat Mass Transfer **56**, 495–500 (2013)
60. Martin, P., Aksamija, Z., Pop, E., Ravaioli, U.: Impact of phonon-surface roughness scattering on thermal conductivity of thin Si nanowires. Phys. Rev. Lett. **102**, 125503 (2009)
61. Mattia, D., Calabrò, F.: Explaining high flow rate of water in carbon nanotubes via solid-liquid molecular interactions. Microfluid. Nanofluid. **13**, 125–130 (2012)

62. Maurer, J., Tabeling, P., Joseph, P., Willaine, H.: Second-order slip laws in microchannels for helium and hydrogen. Phys. Fluid. **15**, 2613–2621 (2003)
63. Mitsuya, Y.: Modified Reynolds equation for ultra-thin film gas lubrication using 1.5-order slip-flow model and considering surface accommodation coefficient. J. Tribol. T. ASME **115**, 289–295 (1993)
64. Mongiovì, M.S.: Extended irreversible thermodynamics of liquid helium II. Phys. Rev. B **48**, 6276–6283 (1993)
65. Mongiovì, M.S., Jou, D.: Thermodynamical derivation of a hydrodynamical model of inhomogeneous superfluid turbulence. Phys. Rev. B **75**, 024507 (14 pp.) (2007)
66. Moore, A.L., Saha, S.K., Prasher, R.S., Shi, L.: Phonon backscattering and thermal conductivity suppression in sawtooth nanowires. Appl. Phys. Lett. **93**, 083112 (2008)
67. Müller, I., Ruggeri, T.: Rational Extended Thermodynamics, 2nd edn. Springer, New York (1998)
68. Pershin, Y.V., Di Ventra, M.: Memory effects in complex materials and nanoscale systems. Adv. Phys. **60**, 145–227 (2011)
69. Putterman, S.J.: Superfluid Hydrodynamics. North Holland, Amsterdam (1974)
70. Roldughin, V.Y.: Nonequilibrium thermodynamics of boundary conditions for rarefied gases and related phenomena. Adv. Colloid Interf. Sci. **65**, 1–35 (1996)
71. Saluto, L., Mongiovì, M.S., Jou, D. Longitudinal counterflow in turbulent liquid helium: velocity profile of the normal component. Z. Angew. Math. Phys. **65**, 531–548 (2014)
72. Sciacca, M., Mongiovì, M.S., Jou, D.: A mathematical model of counterflow superfluid turbulence describing heat waves and vortex-density waves. Math. Comput. Model. **48**, 206–221 (2008)
73. Sciacca, M., Sellitto, A., Jou, D.: Transition to ballistic regime for heat transport in helium II. Phys. Lett. A. **378**, 2471–2477 (2014)
74. Sellitto, A., Alvarez, F.X., Jou, D.: Second law of thermodynamics and phonon-boundary conditions in nanowires. J. Appl. Phys. **107**, 064302 (7 pp.) (2010)
75. Sellitto, A., Alvarez, F.X., Jou, D.: Temperature dependence of boundary conditions in phonon hydrodynamics of smooth and rough nanowires. J. Appl. Phys. **107**, 114312 (7 pp.) (2010)
76. Sellitto, A., Alvarez, F.X., Jou, D.: Phonon-wall interactions and frequency-dependent thermal conductivity in nanowires. J. Appl. Phys. **109**, 064317 (8 pp.) (2011)
77. Sellitto, A., Alvarez, F.X., Jou, D.: Geometrical dependence of thermal conductivity in elliptical and rectangular nanowires. Int. J. Heat Mass Transfer **55**, 3114–3120 (2012)
78. Struchtrup, H.: Macroscopic transport equations for rarefied gas flows: approximation methods in kinetic theory. In: Interaction of Mechanics and Mathematics. Springer, New York (2005)
79. Tabeling, P.: Introduction to Microfluidics. Oxford University Press, Oxford (2005)
80. Tešanović, Z., Jarić, M.V., Maekawa, S.: Quantum transport and surface scattering. Phys. Rev. Lett. **57**, 2760–2763 (1986)
81. Torrilhon, M., Struchtrup, H.: Boundary conditions for regularized 13-moment equations for microchannel flows. J. Comp. Phys. **227**, 1982–2011 (2008)
82. Tzou, D.Y.: Nonlocal behavior in phonon transport. Int. J. Heat Mass Transf. **54**, 475–481 (2011)
83. Tzou, D.Y.: Macro- to Microscale Heat Transfer: The Lagging Behaviour, 2nd edn. Wiley, New York (2014)
84. Tzou, D.Y., Guo, Z.-Y.: Nonlocal behavior in thermal lagging. Int. J. Thermal Sci. **49**, 1133–1137 (2010)
85. Van Sciver, S.W.: Helium Cryogenics, 2nd edn. Springer, Berlin (2012)
86. Vázquez, F., Márkus, F.: Size effects on heat transport in small systems: dynamical phase transition from diffusive to ballistic regime. J. Appl. Phys. **105**, 064915 (2009)
87. Volz, S. (ed.): Thermal Nanosystems and Nanomaterials. Topics in Applied Physics. Springer, Berlin (2010)
88. Wang, M., Yang, N., Guo, Z.-Y.: Non-Fourier heat conductions in nanomaterials. J. Appl. Phys. **110**, 064310 (7 pp.) (2011)
89. Wilks, J. The Properties of Liquid and Solid Helium. Clarendon Press, Oxford (1967)

90. Wu, L.: A slip model for rarefied gas flows at arbitrary Knudsen number. Appl. Phys. Lett. **93**, 253103 (2008)
91. Xu, M.: Slip boundary condition of heat flux in Knudsen layers. Proc. R. Soc. A **470**, 20130578 (9 pp.) (2014)
92. Xu, M., Li, X.: The modeling of nanoscale heat conduction by Boltzmann transport equation. Int. J. Heat Mass Transf. **55**, 1905–1910 (2012)
93. Zhang, Z.M.: Nano/Microscale Heat Transfer. McGraw-Hill, New York (2007)
94. Ziman, J.M.: Electrons and Phonons. Oxford University Press, Oxford (2001)

Chapter 4
Mesoscopic Description of Effective Thermal Conductivity in Porous Systems, Nanocomposites and Nanofluids

Nanosystems do not only refer to truly small systems, but also to systems characterized by some internal microstructures giving to them some special mechanical, thermal, electrical and optical properties which are very useful in practical applications. Such structures may be nanopores, or nanoparticles, or several parallel very thin layers. By regulating the main features of such internal structure, as for instance their characteristic size, or their average separation distance and their spatial distribution (in the case of nanopores and nanoparticles), as well as their thickness and the materials of nanolayers, one may control the transport properties of those systems.

When the main geometrical features mentioned above are large as compared with the mean-free path of the heat carriers, the thermo-mechanical properties of the corresponding systems are well known in the framework of the classical Fourier theory. The open frontier is to analyze such properties in the case when the aforementioned features are comparable to (or smaller than) the mean-free path of heat carriers, in which case new frontiers emerge.

In this chapter we describe with some detail the application of the model of phonon hydrodynamics to nanopores materials. We also briefly furnish (that is, with less details) some results from other theories for the thermal conductivity both of nanocomposites and superlattices, and of nanofluids, which are the basis of many outstanding applications.

4.1 Pore-Size Dependence of the Effective Thermal Conductivity

We begin our analysis by considering nanoporous materials, and in particular, porous silicon (pSi).

© Springer International Publishing Switzerland 2016
A. Sellitto et al., *Mesoscopic Theories of Heat Transport in Nanosystems*,
SEMA SIMAI Springer Series 6, DOI 10.1007/978-3-319-27206-1_4

The thermal conductivity of pSi has been found to decrease greatly for increasing porosity [10, 15, 17, 21, 36, 59], getting two or three orders of magnitude lower than that of monocrystalline silicon. These low thermal-conductivity values allow us to use this material as thermal insulator in microsensors and microsystems [26, 64]. Furthermore, optimization of its use in optoelectronic applications, due to its outstanding photoluminescence properties, requires a good knowledge of its thermal properties. Because of these applications, this topic has become of much interest in nanoscale heat transport [14, 63, 66]. The experimental results on pSi show that its thermal conductivity is strongly related to the pore size at a given total porosity [59]. Since an increasing porosity may deteriorate the electron transport properties [4, 37], it would be advantageous if its effective thermal conductivity could be controlled by the volume fractions as well as the characteristic size of the pores. Often the study of thermal conductivity in a porous medium is treated from kinetic theory for phonons, or from molecular simulations.

The simplest theoretical model prescribes that the thermal conductivity of a porous medium depends only on the porosity ϕ [52] (defined as the ratio between the volume of pores and the volume of the hosting medium), according to the equation

$$\lambda_{\text{eff}} = \lambda f (\phi) \tag{4.1}$$

wherein $f (\phi)$ means a positive-defined regular function of the porosity, whose value is smaller than 1. Different models differ from each other in the form of this function [16, 17, 21, 39, 42, 60]. For example, in the so-called percolation model, a possible way of modeling that function is

$$f (\phi) = (1 - \phi)^3 .$$

However, for a given porosity, the value of the radius r of the pores also influences the thermal conductivity, especially for very small values of r. This is logical, because small values of r mean a high number of pores (for a given ϕ) and, therefore, a higher area of the boundary between the pores and the material, thus leading to higher phonon scattering. Of course, not all phonons are dispersed in the same way, but those with phonon wavelength comparable to (or smaller than) the pore radius will scatter more intensely than those with long phonon wavelengths.

A simple way to introduce this effect is to replace λ in Eq. (4.1) by $\lambda (1 + \ell_{\text{bulk}}/d)^{-1}$, with ℓ_{bulk} being the phonon mean-free path in the bulk material of the hosting matrix, and d the average separation of pores. If N is the number of the inner pores and V the volume of the medium, d is of the order of

$$d \approx \sqrt[3]{\frac{V}{N}} = \left(\sqrt[3]{\frac{4}{3}} \right) \frac{1}{\sqrt[3]{\phi}} r.$$

This leads to

$$\lambda_{\text{eff}} = \lambda \left[1 + \left(\sqrt[3]{\frac{3\phi}{4}} \right) \frac{\ell_{\text{bulk}}}{r} \right]^{-1} (1 - \phi)^3$$

which is similar to the proposal by Sumirat et al. [61]. Another proposal is that by Lysenko et al. [40], namely,

$$\lambda_{\text{eff}} = \lambda \left(1 + \frac{4}{3} \frac{\ell_{\text{bulk}}}{r} \right)^{-1} (1 - \phi)^3 .$$

In the present section we study the thermal conductivity of pSi from a more thermodynamic perspective, by applying the phonon hydrodynamics to the analysis of the thermal conductivity of porous silicon, considered as a solid matrix with the inclusion of small insulating spheres (see Fig. 4.1 for a qualitative sketch).

Our aim is to explore the influence of the pores size on the effective thermal conductivity λ_{eff}. To achieve that task, it is possible to take advantage from classical fluid-dynamic results [2, 57]. In fact, we previously observed that on a mesoscopic level the heat carriers behave as moving fluid particles, and in some occasions there is a very close relation between the equations of phonon hydrodynamics and those of classical hydrodynamics. This allowed us to establish in Sect. 3.1 a sort of parallelism between the fluid-dynamic quantities and the thermal ones.

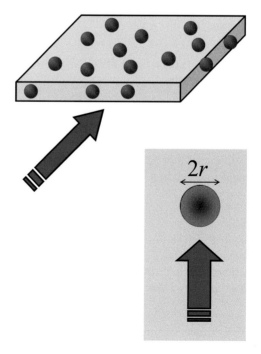

Fig. 4.1 Sketch of a pSi sample. The pores are considered as insulating spheres dispersed in a silicon solid matrix. The pores may be randomly distributed in the matrix (as in the picture), or have a periodic distribution (simple-cubic distribution, or a body-centered cubic distribution). For the sake of simplicity, we suppose that all pores have the same radius equal to r. The dimensions of the pores have been emphasized, in order to see easily their presence. The *arrow* stands for the heat flux

4.1.1   Results from Classical Fluid-Dynamics: The Correction Factors

Phonon-hydrodynamic approach allows us to take advantage from the classical fluid-dynamics to analyze heat-transport problems in nanosystems. In the special case of pSi, to account for the problem of the finite and non-vanishing thermal resistance of the non-conducting pores in the silicon hosting matrix, here we start to model them as rigid spheres, with radius r, moving with a given speed $\mathbf{v}$ in a viscous fluid with a shear viscosity η. In fluid-dynamics, whenever the Reynolds number $\mathrm{Re} = vr/v$ (v being the kinematic viscosity) is smaller enough, the consequent drag-force $\mathbf{D}_\eta$, acting on the external surface of the sphere can be estimated by the Stokes formula, i.e.,

$$\mathbf{D}_\eta = 6\pi \eta r \mathbf{v}. \tag{4.2}$$

We note that the dependence of $\mathbf{D}_\eta$ on $\mathbf{v}$ becomes quadratic when $\mathbf{v}$ gets sufficiently high values. However, for the sake of simplicity, in the next only the linear dependence will be considered.

When the flow becomes rarefied, the effects of a slip flow on the external surface of the sphere must be also taken into account. This means that the aforementioned drag force is reduced. In particular, when the sphere size is smaller than (or comparable to) the molecular mean-free path, Eq. (4.2) has to be modified as follows

$$\mathbf{D}_\eta = 6\pi \eta r \mathbf{v} / \Gamma_1 \tag{4.3}$$

wherein

$$\Gamma_1 = 1 + 2\frac{l}{r}\left(1.257 + 0.4e^{-1.1r/l}\right) \tag{4.4}$$

is the so-called Cunningham correction-factor [18]. The presence of the molecular mean-free path l in Eq. (4.4) accounts for the rarefaction effects: the higher l, the smaller $\mathbf{D}_\eta$. In the literature, other different proposals for this factor can be found [5, 44].

In the case of several rigid spheres adsorbed in the gas flow, the drag on each sphere is given by [25, 29, 31, 53–55]

$$\mathbf{D}_\eta = 6\pi \eta r \mathbf{v} / (\Gamma_1 \Gamma_2) \tag{4.5}$$

where Γ_2 is a further correction factor. Considering a finite volume completely made of gas and spheres, Γ_2 depends on the distribution of the spheres and changes for a simple-cubic distribution (SCD), or a body-centered cubic distribution (BCCD). Denoting by φ the volume-fraction corresponding to the spheres, the correction factor Γ_2 in those situations is given by

$$\Gamma_2 = 1 - 1.76\sqrt[3]{\varphi} + \varphi \quad \text{(SCD)} \tag{4.6a}$$

$$\Gamma_2 = 1 - 1.79\sqrt[3]{\varphi} + \varphi \quad \text{(BCCD).} \tag{4.6b}$$

Furthermore, a random distribution of spheres (RD) is also possible, and in such a case the correction factor Γ_2 is given by the Brinkman expression [12]

$$\Gamma_2 = \left(1 + \frac{3}{\sqrt{2}}\sqrt{\varphi}\right)^{-1} \quad \text{(RD).} \tag{4.7}$$

4.1.2 Theoretical Thermal Conductivity of Porous Silicon

For the heat-transport problem, previous results may be used as starting point to estimate the thermal conductivity of pSi. In fact, suppose that a given amount of heat is flowing through pSi, regarded as a silicon solid matrix with inclusion of small insulating spheres of radius r [2, 57]. Then, the flow of phonons is hindered and reduced by the insulating spheres. Therefore, a nonstandard thermal-drag force $\mathbf{T}_p$, due to the porosity, is acting on each single insulating sphere. In the phonon-hydrodynamic approach, for moderate values of $\mathbf{q}$, we may use the results above to estimate this force. Once the molecular mean-free path l is replaced by the phonon mean-free path ℓ_p, as well as φ is replaced by ϕ, we get for the nonstandard thermal-drag force the following expression

$$\mathbf{T}_p = \frac{6\pi r \ell_p^2}{\Gamma_1 \Gamma_2 \lambda_0}\mathbf{q} \tag{4.8}$$

where Γ_1 is given by Eq. (4.4) and Γ_2 either by one of Eqs. (4.6), or by Eq. (4.7), depending on the distribution of the inner pores. In the presence of a flow through a medium, the standard thermal-drag force

$$\mathbf{T}_s = \frac{V}{\lambda f(\phi)}\mathbf{q} \tag{4.9}$$

being V the volume of the system, has to be further taken into account as well. Thus, if N is the number of the insulating spheres, the total thermal-drag force is

$$\mathbf{T}_r = N\mathbf{T}_p + \mathbf{T}_s.$$

Since $\mathbf{T}_r$ balances the total driving thermal-force $V\nabla T$, due to the applied temperature gradient, one has

$$\left[\frac{6\pi N\ell_p^2 r}{\Gamma_1\Gamma_2} + \frac{V}{f(\phi)}\right]\frac{\mathbf{q}}{\lambda} = V\nabla T. \tag{4.10}$$

Note that we assumed that all pores have the same radius: in this way in Eq. (4.10) we have expressed N in terms of the porosity ϕ, as $N = V/(\phi V_s)$, where $V_s = (4/3)\pi r^3$ is the volume of a single sphere. Then, straightforward calculations allow to obtain the following effective thermal conductivity [2]

$$\lambda_{\text{eff}} = \frac{|\mathbf{Q}|}{A\,|\nabla T|} = \frac{\lambda}{\dfrac{1}{f(\phi)} + \dfrac{9}{2}\phi\dfrac{\text{Kn}^2}{\Gamma_1\Gamma_2}} \tag{4.11}$$

where we have considered that $\mathbf{Q} = A\mathbf{q}$ is the total heat flux, A being the transversal area (perpendicular to the direction of propagation of the heat flux), and $\text{Kn} = \ell_p/r$. Looking at the denominator of Eq. (4.11), it is possible to distinguish clearly two different contributions: the first one related to $f(\phi)$ which accounts for the porosity, and the second one related to the Knudsen number Kn which is related, instead, to the characteristic size of the pores. For low values of Kn Eq. (4.11) reduces to Eq. (4.1).

To compare the theoretical results following from Eq. (4.11) with experimental observations, in Table 4.1 we report experimental data for the thermal conductivity of pSi at the room temperature, for different porosity and pores' radii. In the same table, also the theoretical results, following from the relation $\lambda_{\text{eff}} = \lambda(1-\phi)^3$, are quoted. This way the importance of accounting for the role of the pores radius in the theoretical predictions may be enlightened.

In Table 4.2, instead, we show the theoretical results predicted by Eq. (4.11) for different internal distributions of the pores and for the same cases as in Table 4.1.

As it is possible to see, the theoretical proposal in Eq. (4.11) has a better agreement with experimental observation with respect to Eq. (4.1).

Table 4.1 Experimental data on the thermal conductivity (W/mK) of porous Si for different porosities ϕ (%) and pores' radius r (nm) at $T = 300\,\text{K}$

Source	ϕ	r	λ_{eff} Experimental	λ_{eff} Theoretical
Ref. [10]	60	10	2–5	9.47
Ref. [21]	64	2	0.20	6.91
Ref. [21]	71	2	0.14	3.61
Ref. [21]	79	3	0.06	1.37
Ref. [21]	89	5	0.04	0.19

The theoretical data following from the relation $\lambda_{\text{eff}} = \lambda(1-\phi)^3$ are quoted, too

Table 4.2 Theoretical results on the thermal conductivity (W/mK) of pSi obtained from Eq. (4.11)

Source	ϕ	r	λ_{eff} Eq. (4.11)—SCD	λ_{eff} Eq. (4.11)—BCCD	λ_{eff} Eq. (4.11)—RD
Ref. [10]	60	10	3.41	2.89	6.13
Ref. [21]	64	2	0.92	0.73	2.17
Ref. [21]	71	2	0.83	0.71	1.56
Ref. [21]	79	3	0.74	0.69	0.98
Ref. [21]	89	5	0.18	0.18	0.18

Different internal distributions of the pores (SCD, BCCD and RD), different porosities ϕ (%) and pores' radius r (nm) have been analyzed. The sample is supposed at 300 K

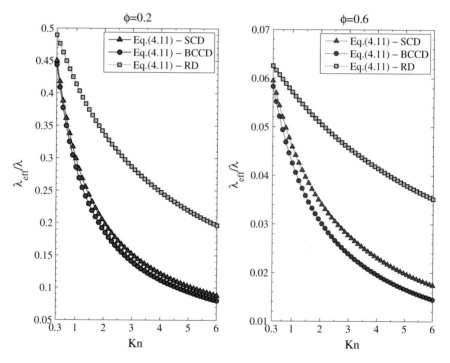

Fig. 4.2 Behavior of the ratio between λ_{eff} and λ as a function of Kn at $T = 300\,\text{K}$ arising from Eq. (4.11). Two different values of the porosity have been chosen, namely, $\phi = 0.2$ and $\phi = 0.6$. Moreover, for each value of ϕ, the SCD, BCCD and RD of the pores have been considered. Different length-scales have been used for both situations

In Fig. 4.2 we also plot the behavior of the ratio of λ_{eff} to λ at the room temperature and for two given values of porosity (i.e., $\phi = 0.2$ and $\phi = 0.6$), as a function of Kn, in the cases of SCD, BCCD and RD of the pores.

Figure 4.2 allows to conclude that for increasing Kn the effective thermal conductivity of pSi decreases whatever the internal distribution of the pores is. In

Fig. 4.3 Pore with rough walls. The roughness may be described by the roughness height (Δ), and the separation of neighboring roughness peaks (D). Whenever a particle hits a peak, it may be reflected backward. Otherwise it is reflected in a diffusive or in a specular way

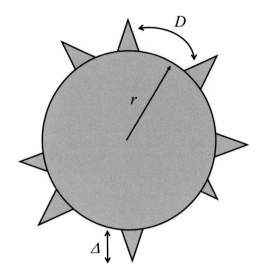

particular, the reduction in λ_{eff} seems to be rather strong, and only partly reduced by a random distribution of the pores.

Both from Table 4.1 and Fig. 4.2 it follows that, for a given porosity, the BCCD of the internal pores seems to be the best combination whenever pSi is used for device isolation in integrated circuits, or for heat protection of artificial satellite during atmospheric reentry.

As final remark let's recall that Eq. (4.4) is valid in the case of a smooth sphere, with diffusive particles-wall collisions. When the radius of the sphere is comparable to the particles mean-free path, the wall roughness, as well as the wall heat flux, may be especially relevant [57] (see Fig. 4.3 for a sketch of the roughness of the pore wall).

If the assumptions (3.40) are made about the coefficients C and α, the correction factor Γ_1 corresponding to this new situation may be calculated [5]. As result, when $r \approx \ell_p$ it is no longer given by Eq. (4.4), but by

$$\Gamma_1 = \frac{1 + 2C\dfrac{\ell_p}{r}}{1 + 3C\dfrac{\ell_p}{r} + \dfrac{3\alpha}{\pi}\dfrac{\ell_p^2}{r^2}} \qquad (4.12)$$

where C and α are now suitable coefficients which may depend both on the radius of the pores, and on the roughness of the pores [57]. From the practical point of view, Eq. (4.12) cannot be used for the correction factor Γ_1 in order to evaluate the corresponding effective thermal conductivity, due to the lacking of experimental data for rough-walled pSi. However, the increasing interest in pSi is fostering experiments on its thermal properties. Thus, the knowledge in this area is in progress, and Eq. (4.12) may constitute the starting point of future research.

4.1.3 An EIT Approach

Other interesting mesoscopic proposals for modeling the thermal conductivity in porous media can be found in literature [27, 40–42, 61]. In Refs. [41, 42], in particular, the effect of the presence of nanopores in a hosting matrix on the thermal conductivity of silicon is investigated by referring to the thermodynamic formulation of EIT, coupled to the so-called effective medium approximation. In such approximation which we will comment in more detail below, without taking into account the matrix-particle boundary resistance, the effective thermal conductivity of a matrix of material m with dispersed nanoparticles of material p is [41, 42]

$$\lambda_{\text{eff}} = \lambda_m \left[\frac{2\lambda_m + \lambda_p + 2\phi \left(\lambda_p - \lambda_m\right)}{2\lambda_m + \lambda_p - \phi \left(\lambda_p - \lambda_m\right)} \right] \qquad (4.13)$$

with λ_m being the thermal conductivity of the hosting matrix (which coincides with the bulk value), and λ_p is the thermal conductivity of the nanoparticles. In Refs. [41, 42] it has been proposed to assume that the pores have air inside, in such a way that λ_p would be the thermal conductivity of air. To bring into Eq. (4.13) the influence of the pore radius, it has been proposed to use for λ_p an effective thermal conductivity of the form of that introduced in Sect. 1.3 and obtained from the continued-fraction expansion [19, 23] of the thermal conductivity in terms of the Knudsen number, as obtained by using higher-order fluxes in EIT formalism. Thus, Machrafi and Lebon proposed to replace in Eq. (4.13) the thermal conductivity λ_p with

$$\lambda_{\text{eff},p} = \frac{3\lambda_p}{4\pi^2 \, \text{Kn}^2} \left[\frac{2\pi \, \text{Kn}}{\arctan\left(2\pi \, \text{Kn}\right)} - 1 \right]. \qquad (4.14)$$

By coupling Eqs. (4.13) and (4.14), it was achieved a satisfactory agreement with experimental results. This shows how mesoscopic approach provide different alternative ways to deal with the effective thermal conductivity of nanoporous materials.

4.1.4 Microporous Films

An analogous analysis could be made for microporous films with cylindrical holes in them [59]. Such systems are especially interesting when the holes are distributed in an ordered periodical way. In such a case one has the so-called phononic crystals, in which one may achieve selective effects on some given phonon frequencies because of resonances caused by destructive interference. Such special effects are not strictly describable if it is assumed that the mean-free path does not depend on frequency.

4.2 Nanocomposites, Superlattices and Nanofluids

Besides nanoporous materials, other two kinds of systems of much interest are nanocomposites and nanofluids, namely, systems which are composed of a heat conducting matrix (solid in the former case, and liquid in the latter case) and a dispersion of heat conducting nanoparticles. Depending on the radius, size, number density and spatial distribution of the nanoparticles (in the case of nanocomposites), one may achieve a considerable degree of control on the transport properties of the system, either of heat, electric or thermoelectric transport coefficients.

Some of the most well-known nanocomposites are superlattices [1], in which the particles are distributed in a spatially ordered pattern, and the nanoparticles may be so small as to behave as quantum dots [3, 6, 34, 56, 62]. Yet another possibility is to organize the system as an ensemble of parallel thin layers of alternating materials. In this case, there are two sources of reduction of thermal conductivity: that due to ballistic behavior of phonons inside each layer (if they are thin enough to be comparable to the phonon mean-free path), and that due to thermal resistance in the interfaces. Furthermore, in metals and semiconductors, where electrons and holes also contribute to thermal conductivity, superlattices may be designed in such a way to strongly reduce phonon transport without reducing electron transport, because of the widely different values of phonon and electron wavelengths. This is especially relevant in enhancing the efficiency of thermoelectric energy conversion, which requires to reduce as much as possible the thermal conductivity. Since the latter is directly related to electronic thermal conductivity, the reduction of phonon thermal conductivity (but not of the electrical one) is a way to increase such efficiency. Superlattices made of silicon and germanium [13, 38, 50], or of indium arsenide and gallium antimonide [49], or nanocomposites made of silicon dioxide or aluminium nitride particles embedded in epoxy matrix [7, 30] are especially well-known.

Another interesting aspect of nanocomposites may be graded systems [11], in which the composition of the system is not homogeneous, but changes along some directions. These systems are increasingly used in technology [28, 32, 33, 46]. One of their applications is in heat diodes, to achieve some degree of rectification. We will deal a little bit with this topic in Sect. 5.2 of Chap. 5. Further relevant application of nanocomposites is in thermoelectricity. For instance, the alloys $Bi_{2-2x} Te_{3-2x} Pb_x$, or $(Bi_{1-x} Sb_x)_2 Te_3$, with the stoichiometric index x changing from 0 to 1, are of interest in thermoelectric energy generators [58]. Indeed, as it will be seen in Chap. 6, the efficiency of thermoelectric energy conversion depends on the nondimensional product ZT, with Z being the figure-of-merit. For $Bi_2 Te_3$ and $Pb Te$ (two of the best thermoelectric materials), ZT is maximum around 370 K and 670 K, respectively. Thus, to maximize that product along a system between 670 and 370 K is better to use $Pb Te$ at the hot side, $Bi_2 Te_3$ at the cold side, and $Bi_{2-2x} Te_{3-2x} Pb_x$ in the middle (with x changing from 1 in the hot side to 0 in the cold side).

Here we briefly sketch some simple ways which have been proposed to deal with the thermal conductivity of these systems.

4.2.1 Thermal Conductivity of Nanocomposites

A Nanocomposite is a multi-phase solid material wherein one of the phases has a characteristic size of the order of tens nanometers, or structures having nanoscale repeat distances between the different phases that make up the material. For the sake of simplicity, in the next we assume that the inner particles have a spherical shape and are dispersed in a homogeneous matrix.

If the dimensions of the spheres is larger than the phonon mean-free path, then the heat conduction is well described by the usual Fourier law with the following effective thermal conductivity [24, 47]

$$\lambda_{\text{eff}} = \lambda_m \left\{ \frac{2\lambda_m + (1+2Y)\lambda_p + 2\phi\left[(1-Y)\lambda_p - \lambda_m\right]}{2\lambda_m + (1+2Y)\lambda_p - \phi\left[(1-Y)\lambda_p - \lambda_m\right]} \right\} \tag{4.15}$$

wherein ϕ is the volume fraction of the dispersed particles, λ_m and λ_p are the thermal conductivity of the hosting matrix and of the inner particles, respectively, and Y is a dimensionless parameter which accounts for the particles-matrix interface. That parameter may be expressed as $Y = a_p / (C_B \lambda_m)$, with a_p being the radius of the spherical particles, and C_B the thermal boundary resistance coefficient between the matrix and the particles. It is easy to see that whenever $Y \equiv 0$, Eq. (4.15) reduces to Eq. (4.13).

Whenever the characteristic dimension of the inner spheres is smaller than the phonon mean-free path, Eq. (4.15) is no longer valid [45], and the heat transport in nanocomposites has to be analyzed with the Boltzmann-Peierls theory wherein the heat is seen as a gas of phonons [7, 20, 22, 51]. From the phenomenological point of view, in order to extend Eq. (4.15) to nanocomposites, Minnich and Chen [45] proposed to replace the parameters λ_m, λ_p and Y therein with the following effective coefficients

$$\lambda_{m,\text{eff}} = \frac{\lambda_m}{1 + \dfrac{\ell_{m,b}}{\ell_{m,\text{coll}}}} = \frac{\lambda_m}{1 + \dfrac{3\phi\ell_m}{4a_p}} \tag{4.16a}$$

$$\lambda_{p,\text{eff}} = \frac{\lambda_p}{1 + \dfrac{\ell_{p,b}}{\ell_{p,\text{coll}}}} = \frac{\lambda_p}{1 + \dfrac{i\ell_p}{2a_p}} \tag{4.16b}$$

$$Y = 4 \left(\frac{c_{v,m}\overline{\mathbf{v}}_m + c_{v,p}\overline{\mathbf{v}}_p}{c_{v,m}\overline{\mathbf{v}}_m c_{v,p}\overline{\mathbf{v}}_p} \right) \tag{4.16c}$$

wherein $\ell_{m,b}$ and $\ell_{p,b}$ are the phonon mean-free path in the bulk material of the matrix and of the particles, respectively, $\ell_{m,\text{coll}}$ and $\ell_{p,\text{coll}}$ are the collision phonon mean-free path in the matrix and in the particles, respectively, $c_{v,m}$ and $c_{v,p}$ are the volumetric specific heats of the bulk material of the matrix and of the particles, and $\bar{v}_m$ and $\bar{v}_p$ are the average phonon speed in the matrix and in the particles.

In Ref. [7] a third route, which is a hybrid of the two approaches above, is followed. In that reference a new consideration of the matrix-particles interface is made, and Eq. (4.15) is generalized as

$$\lambda_{\text{eff}} = \hat{g}\lambda^s_{\text{eff}} + (1 - \hat{g})\,\lambda^d_{\text{eff}} \tag{4.17}$$

namely, the effective thermal conductivity of nanocomposites is a linear combination of λ^s_{eff} (corresponding to the case when all phonon-interface interactions are specular) and λ^d_{eff} (corresponding to the case when all phonon-interface interactions are diffuse). In Eq. (4.17) $\hat{g}$ is a phenomenological parameter (such that $0 \leq \hat{g} \leq 1$) having the physical interpretation of the probability of the specular scattering of phonons on the particle-matrix interface [7]. The explicit expression for λ_{eff} in Eq. (4.17) is very long, owing to the very complex forms of λ^s_{eff} and λ^d_{eff} [see Eq. (8) in Ref. [7]] which involve several parameters. Therefore, for the sake of simplicity, we do not report here that formula.

4.2.2   Thermal Conductivity of Superlattices

Several expressions for the thermal conductivity of layered superlattices have been proposed, either for the cross-plane (CP) direction, or the in-plane (IP) direction (see Fig. 4.4 for a sketch of a superlattice), depending on whether heat flows orthogonally, or parallel to the layers.

As a particular example, in Ref. [1] the following expressions for the thermal conductivity of a superlattice have been derived:

$$\lambda_{\text{CP}} = \frac{\lambda_{1,\text{eff}}L_1 + \lambda_{2,\text{eff}}L_2}{L_1 + L_2} \tag{4.18a}$$

$$\lambda_{\text{IP}} = \frac{L_1 + L_2}{\dfrac{L_1}{\lambda_{1,\text{eff}}} + \dfrac{L_2}{\lambda_{2,\text{eff}}}} \tag{4.18b}$$

wherein $L_i = n_i d_i$ $(i = 1, 2)$ with n_i being the number of the layers of material i, and d_i their thickness, whereas $\lambda_{i,\text{eff}}$ $(i = 1, 2)$ means the effective thermal conductivity of the thin layer which is given as in Eq. (1.26), that is,

$$\lambda_{i,\text{eff}} = \frac{\lambda_i d_i^2}{2\pi^2 \ell_{p,i}^2}\left[\sqrt{1 + \left(\frac{2\pi \ell_{p,i}}{d_i}\right)^2} - 1\right] \tag{4.19}$$

with $\ell_{p,i}$ being the phonon mean-free path in the layer of material i.

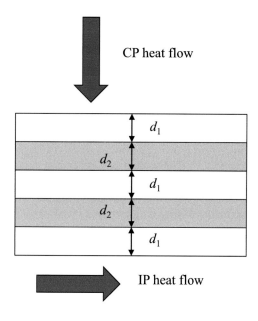

Fig. 4.4 Sketch of a superlattice with representation of cross plane (CP) and in plane (IP) heat flows

The effect of the thermal boundary resistance (R) between neighboring layers in the CP can be taken into account by considering the following thermal conductivity:

$$\lambda_{CP+R} = \frac{\lambda_{CP}}{1 + \dfrac{2\lambda_{CP}R_M}{\ell_{p,1} + \ell_{p,2}}} \quad (4.20)$$

wherein $R_M = (R_{12} + R_{21})/2$, with $R_{hk} = R'_{hk} - \ell_{p,1}/\lambda_1 - \ell_{p,2}/\lambda_2$, and R'_{hk} is a transmission coefficient of heat from h to k material [see Eq. (8) in Ref. [1] for its expression]. In these expressions it is shown the influence of the thickness of the layers - as considered in Eq. (4.19) - the number of layers - as related to L_1 and L_2 in Eq. (4.18) - and, in the case of cross-plane transport, the role of the thermal resistance at interfaces. Such a thermal resistance is by itself an interesting topic in nano heat transfer, which may be described from acoustic mismatch models, or diffusive mismatch models, or a combination of both. Here we only take it as a coefficient for the sake illustration.

4.2.3 Thermal Conductivity of Nanofluids

The thermal aspects of fluids are especially relevant in refrigeration and energy storage. In these applications, the specific and latent heats, the boiling temperature and the freezing one, the thermal conductivity, the chemical stability, and some other properties are important in the selection and the use of the fluids. In particular, adding nanoparticles to the fluids provides a way of modifying and controlling such properties, as for example their thermal conductivity (which rests the main topic of this book).

The common base fluids usually include water, ethylene glycol and oil, whereas the nanoparticles may be made of metals, oxides, carbides, or carbon nanotubes. Nanofluids show relevant novel properties, the main of whose is the enhanced thermal conductivity. This makes nanofluids very appealing for many applications in heat transfer, including microelectronics, fuel cells, pharmaceutical processes, and hybrid-powered engines, engine cooling/vehicle thermal management, domestic refrigerator, chiller, heat exchanger, and many others.

The thermal conductivity of nanofluids is related to the nature and properties both of the base fluid, and of the suspended particles, as well as their volume fraction and the nature of the particle-matrix interface [35].

When the nanoparticles are completely dispersed in the hosting fluid and have spherical shapes, a possible way to model the thermal conductivity $\lambda_{\rm nf}$ of nanofluids is to use the Maxwell model [43] for the electrical conductivity of rigid particles in a fluid, namely,

$$\lambda_{\rm nf} = \lambda_f \frac{\overline{\alpha} + 2 - 2\phi\,(1 - \overline{\alpha})}{\overline{\alpha} + 2 + \phi\,(1 - \overline{\alpha})} \tag{4.21}$$

wherein λ_f is the thermal conductivity of the fluid, and $\overline{\alpha} = \lambda_n/\lambda_f$ with λ_n being the thermal conductivity of the nanoparticles. We note that Eq. (4.21) is essentially Eq. (4.13) (or Eq. (4.15) with $Y \equiv 0$). Whenever the suspended particles are not spherical, Eq. (4.21) should be replaced by [35]

$$\lambda_{\rm nf} = \lambda_f \frac{\overline{\alpha} + n - 1 - (n - 1)\,\phi\,(1 - \overline{\alpha})}{\overline{\alpha} + n - 1 + \phi\,(1 - \overline{\alpha})} \tag{4.22}$$

where n is a shape coefficient which is equal to 3 in the case of spherical nanoparticles,[1] and equal to 6 in the case of cylindrical nanoparticles.

In the case of fibers of high aspect ratio, other authors proposed

$$\lambda_{\rm nf} \approx \lambda_f + \frac{\phi\lambda_p}{3} \tag{4.23}$$

when the nanoparticles are made fibers of high aspect ratio [48], or when they are carbon nanotubes completely straight (i.e., without any curvature or wrapping) [65]. When the nanoparticles are carbon nanotubes with many curves and wrap around themselves (i.e., they form a rope-ball shape), one has [65]

$$\lambda_{\rm nf} \approx \lambda_f + 3\phi\lambda_p. \tag{4.24}$$

---

[1]In this case Eqs. (4.21) and (4.22) coincide.

In Refs. [8, 9], a model for the thermal conductivity of nanofluids has been also proposed to account for the different interactions between the nanoparticles (assumed to be rigid spheres) and the matrix interface.

Thus, though the ideas used in the analysis of thermal conductivity of nanofluids are similar to those of nanocomposites, a wider diversity of the geometry of the suspended particles is considered. Indeed, fluids are less restrictive than solids in accommodating internal particles. One further problem which appears in nanofluids is related to the density of the particles, which should be similar to that of the hosting fluid, in order to avoid the effects of a fast sedimentation, leading to a separation of the fluid and the particles.

Let us finally comment that neither in nanocomposites, nor in nanofluids the thermal conductivity is the only physical parameter to be taken into account. In solid nanocomposites, electric conductivity, Seebeck coefficient, and mechanical properties should be considered. In fluids, the suspended particles increase the value of the viscosity and give to it non-Newtonian features. Thus, a full consideration of the problem is required to maximize some properties of the system, as for instance, the efficiency of thermoelectric energy conversion, or the accumulation, transport and release of thermal energy in fluids.

References

1. Alvarez, F.X., Alvarez-Quintana, J., Jou, D., Rodriguez-Viejo, J. Analytical expression for thermal conductivity of superlattices. J. Appl. Phys. **107**, 084303 (8 pp.) (2010)
2. Alvarez, F.X., Jou, D., Sellitto, A.: Pore-size dependence of the thermal conductivity of porous silicon: a phonon hydrodynamic approach. Appl. Phys. Lett. **97**, 033103 (3 pp.) (2010)
3. Alvarez-Quintana, J., Alvarez, F.X., Rodriguez-Viejo, J., Jou, D., Lacharmoise, P.D., Bernardi, A., Goñi, A.R., Alonso, M.I.: Cross-plane thermal conductivity reduction of vertically uncorrelated Ge/Si quantum dot superlattices. Appl. Phys. Lett. **93**, 013112 (3 pp.) (2008)
4. Aroutiounian, V.M., Ghulinyan, M.Z.: Electrical conductivity mechanisms in porous silicon. Phys. Status Solidi A **197**, 462–466 (2003)
5. Bailey, C.L., Barber, R.W., Emerson, D.R., Lockerby, D.A., Reese, J.M.: A critical review of the drag force on a sphere in the transition flow regime. AIP Conf. Proc. **762**, 743–748 (2005)
6. Balandin, A.A., Lazarenkova, O.L.: Mechanism for thermoelectric figure-of-merit enhancement in regimented quantum dot superlattices. Nat. Mater. **82**, 415 (3 pp.) (2003)
7. Behrang, A., Grmela, M., Dubois, C., Turenne, S., Lafleur, P.G.: Influence of particle-matrix interface, temperature, and agglomeration on heat conduction in dispersions. J. Appl. Phys. **114**, 014305 (9 pp.) (2013)
8. Behrang, A., Grmela, M., Dubois, C., Turenne, S., Lafleur, P.G., Lebon, G.: Effective heat conduction in dispersion of wires. Appl. Phys. Lett. **104**, 063106 (4 pp.) (2014)
9. Behrang, A., Grmela, M., Dubois, C., Turenne, S., Lafleur, P.G., Lebon, G.: Effective heat conduction in hybrid sphere & wire nanodispersions. Appl. Phys. Lett. **104**, 233111 (5 pp.) (2014)
10. Benedetto, G., Boarino, L., Spagnolo, R.: Evaluation of thermal conductivity of porous silicon layers by a photoacoustic method. Appl. Phys. A Mater. Sci. Process. **64**, 155–159 (1997)
11. Birman, V., Byrd, L.W.: Modeling and analysis of functionally graded materials and structures. Appl. Mech. Rev. **60**, 197–216 (2007)

12. Brinkman, H.C.: A calculation of the viscous force exerted by a flowing fluid on a dense swarm of particles. Appl. Sci. Res. **A1**, 27–34 (1947)
13. Brunner, K.: Si/Ge nanostructures. Rep. Prog. Phys. **65**, 27–72 (2002)
14. Chen, G.: Nanoscale Energy Transport and Conversion - A Parallel Treatment of Electrons, Molecules, Phonons, and Photons. Oxford University Press, Oxford (2005)
15. Chung, J.D., Kaviany, M.: Effects of phonon pore scattering and pore randomness on effective conductivity of porous silicon. Int. J. Heat Mass Transf. **43**, 521–538 (2000)
16. Criado-Sancho, J.M., del Castillo, L.F., Casas-Vázquez, J., Jou, D.: Theoretical analysis of thermal rectification in a bulk Si/nanoporous Si device. Phys. Lett. A **19**, 1641–1644 (2012)
17. Criado-Sancho, J.M., Alvarez, F.X., Jou, D.: Thermal rectification in inhomogeneous nanoporous Si devices. J. Appl. Phys. **114**, 053512 (2013)
18. Cunningham, E.: On the velocity of steady fall of spherical particles through fluid medium. Proc. R. Soc. Lond. A **83**, 357–365 (1910)
19. Dedeurwaerdere, T., Casas-Vázquez, J., Jou, D., Lebon, G.: Foundations and applications of a mesoscopic thermodynamic theory of fast phenomena. Phys. Rev. E **53**, 498–506 (1996)
20. Dreyer, W., Struchtrup, H.: Heat pulse experiments revisited. Contin. Mech. Thermodyn. **5**, 3–50 (1993)
21. Gesele, G., Linsmeier, J., Drach, V., Fricke, J., Arens-Fischer, R.: Temperature-dependent thermal conductivity of porous silicon. J. Phys. D Appl. Phys. **20**, 2911–2916 (1997)
22. Grmela, M., Lebon, G., Dauby, P.C.: Multiscale thermodynamics and mechanics of heat. Phys. Rev. E **83**, 061134 (15 pp.) (2011)
23. Győry, E., Márkus, F.: Size dependent thermal conductivity in nano-systems. Thin Solid Films **565**, 89–93 (2014)
24. Hasselman, D.P.H., Johnson, L.F.: On the periodic fundamental solutions of the Stokes equations and their application to viscous flow past a cubic array of spheres. J. Compos. Mater. **21**, 508–515 (1987)
25. Hill, R.J., Koch, D.L., Ladd, A.J.C.: Moderate-Reynolds-number flows in ordered and random arrays of spheres. J. Fluid Mech. **448**, 243–278 (2001)
26. Imai, K.: A new dielectric isolation method using porous silicon. Solid State Electron. **24**, 159–164 (1981)
27. Jean, V., Fumeron, S., Termentzidis, K., Tutashkonko, S., Lacroix, D.: Monte Carlo simulations of phonon transport in nanoporous silicon and germanium. J. Appl. Phys. **115**, 024304 (13 pp.) (2014)
28. Jeffrey Snyder, G., Toberer, E.S.: Complex thermoelectric materials. Nat. Mater. **7**, 105–114 (2008)
29. Kim, J.Y., Yoon, B.J.: The effective conductivities of composites with cubic arrays of spheroids and cubes. J. Compos. Mater. **33**, 1344–1362 (1999)
30. Kochetov, R., Korobko, A.V., Andritsch, T., Morshuis, P.H.F., Picken, S.J., Smit, J.J.: Modelling of the thermal conductivity in polymer nanocomposites and the impact of the interface between filler and matrix. J. Phys. D Appl. Phys. **44**, 395401 (12 pp.) (2011)
31. Kushch, V.I.: Conductivity of a periodic particle composite with transversely isotropic phases. Proc. R. Soc. A **453**, 65–76 (1997)
32. Kuznetsov, V.L.: Functionally graded materials for thermoelectric applications. In: Rowe, D.M. (ed.) Thermoelectrics Handbook: Macro to Nano – Sec. 38. CRC Press, Boca Raton (2005)
33. Kuznetsov, V.L., Kuznetsova, L.A., Kaliazin, A.E., Rowe, D.M.: High performance functionally graded and segmented Bi_2Te_3-based materials for thermoelectric power generation. J. Mater. Sci. **37**, 2893–2897 (2002)
34. Lazarenkova, O.L., Balandin, A.A.: Electron and phonon energy spectra in a three-dimensional regimented quantum dot superlattice. Phys. Rev. B **66**, 245319 (9 pp.) (2002)
35. Lebon, G.: Heat conduction at micro and nanoscales: a review through the prism of Extended Irreversible Thermodynamics. J. Non-Equilib. Thermodyn. **39**, 35–59 (2014)
36. Lee, J.-H., Grossman, J.C., Reed, J., Galli, G.: Lattice thermal conductivity of nanoporous Si: molecular dynamics study. Appl. Phys. Lett. **91**, 223110 (3 pp.) (2007)

37. Lee, H., Vashaee, D., Wang, D.Z., Dresselhaus, M.S., Ren, Z.F., Chen, G.: Effects of nanoscale porosity on thermoelectric properties of SiGe. J. Appl. Phys. **107**, 094308 (7 pp.) (2010)
38. Li, D., Wu, Y., Fan, R., Yang, P., Majumdar, A.: Thermal conductivity of Si/SiGe superlattice nanowires. Appl. Phys. Lett. **83**, 3186 (3 pp.) (2003)
39. Looyenga, H.: Dielectric constants of heterogeneous mixtures. Physica **31**, 401–406 (1965)
40. Lysenko, V., Roussel, P., Remaki, B., Delhomme, G., Dittmar, A., Barbier, D., Strikha, V., Martelet, C.: Study of nano-porous silicon with low thermal conductivity as thermal insulating material. J. Porous Mater. **7**, 177–182 (2000)
41. Machrafi, H., Lebon, G.: Effective thermal conductivity of spherical particulate nanocomposites: comparison with theoretical models, Monte Carlo simulations and experiments. Int. J. Nanosci. **13**, 1450022 (7 pp.) (2014)
42. Machrafi, H., Lebon, G.: Size and porosity effects on thermal conductivity of nanoporous material with an extension to nanoporous particles embedded in a host matrix. Phys. Lett. A **379**, 968–973 (2015)
43. Maxwell, J.C.: A Treatise on Electricity and Magnetism, 2nd edn. Oxford University Press, Cambridge (1904)
44. Millikan, R.A.: The general law of fall of a small spherical body through a gas, and its bearing upon the nature of molecular reflection from surfaces. Phys. Rev. **22**, 1–23 (1923)
45. Minnich, A., Chen, G.: Modified effective medium formulation for the thermal conductivity of nanocomposites. Appl. Phys. Lett. **91**, 073105 (3 pp.) (2007)
46. Müller, E., Drašar, Č., Schilz, J., Kaysser, W.A.: Functionally graded materials for sensor and energy applications. Mater. Sci. Eng. A**362**, 17–39 (2003)
47. Nan, C.-W., Birringer, R., Clarke, D.R., Gleiter, H.: Effective thermal conductivity of particulate composites with interfacial thermal resistance. J. Appl. Phys. **81**, 6692–6699 (1997)
48. Nan, C.-W., Shi, Z., Lin, Y.: A simple model for thermal conductivity of carbon nanotube-based composites. Chem. Phys. Lett. **375**, 666–669 (2003)
49. Nguyen, B.-M., Hoffman, D., Delaunay, P.-Y., Razeghi, M.: Dark current suppression in type II InAs/GaSb superlattice long wavelength infrared photodiodes with M-structure barrier. Appl. Phys. Lett. **91**, 163511 (1 p.) (2007)
50. Pchelyakov, O.P., Bolkhovityanov, Y.B., Dvurechenskiĭ, A.V., Sokolov, L.V., Nikiforov, A.I., Yakimov, A.I., Voigtländer, B.: Silicon-germanium nanostructures with quantum dots: formation mechanisms and electrical properties. Semiconductors **34**, 1229–1247 (2000)
51. Reissland, J.A.: The Physics of Phonons. Wiley, London (1973)
52. Russell, H.W.: Principles of heat flow in porous insulators. J. Am. Ceram. Soc. **18**, 1–5 (1935)
53. Sangani, A.S., Acrivos, A.: Slow flow through a periodic array of spheres. Int. J. Multiphase Flow **8**, 343–360 (1982)
54. Sangani, A.S., Acrivos, A.: Creeping flow through cubic arrays of spherical bubbles. Int. J. Multiphase Flow **9**, 181–185 (1983)
55. Sangani, A.S., Acrivos, A.: The effective conductivity of a periodic array of spheres. Proc. R. Soc. A **386**, 263–275 (1983)
56. Sellitto, A.: A phonon-hydrodynamic approach to thermal conductivity of Si-Ge quantum dot superlattices. Appl. Math. Model. **39**, 4687–4698 (2015)
57. Sellitto, A., Jou, D., Cimmelli, V.A.: A phenomenological study of pore-size dependent thermal conductivity of porous silicon. Acta Appl. Math. **122**, 435–445 (2012)
58. Snyder, G.J., Toberer, E.S.: Complex thermoelectric materials. Nat. Mater. **7**, 105–114 (2008)
59. Song, D., Chen, G.: Thermal conductivity of periodic microporous silicon films. Appl. Phys. Lett. **84**, 687–690 (2004)
60. Sturm, J., Grosse, P., Theiss, W.: Effective dielectric functions of alkali halide composites and their spectral representation. Z. Phys. B Condens. Matter **83**, 361–365 (1991)
61. Sumirat, I., Ando, Y., Shimamura, S.: Theoretical consideration of the effect of porosity on thermal conductivity of porous materials. J. Porous Mater. **13**, 439–443 (2006)
62. Taylor, T.C.H.P.J., Walsh, M.P., La Forge, B.E.: Quantum dot superlattice thermoelectric materials and devices. Science **297**, 2229–2232 (2002)

63. Tzou, D.Y.: Macro- to Microscale Heat Transfer: The Lagging Behaviour, 2nd edn. Wiley, New York (2014)
64. Wanatabe, Y., Arita, Y., Yokoyama, T., Igarashi, Y.: Formation and properties of porous silicon and its applications. J. Electrochem. Soc. Solid State Sci. Technol. **122**, 1351–1355 (1975)
65. Yu, W., France, D.M., Timofeeva, E.V., Singh, D.: Effective Thermal conductivity models for carbon nanotube-based nanofluids. J. Nanofluids **2**, 69–73 (2013)
66. Zhang, Z.M.: Nano/Microscale Heat Transfer. McGraw-Hill, New York (2007)

Chapter 5
Weakly Nonlocal and Nonlinear Heat Transport

The thermo-mechanical behavior of miniaturized systems, the characteristic lengths of which is of the order of few nanometers, is strongly influenced by memory, nonlocal, and nonlinear effects [1, 18, 27, 50]. In one-dimensional steady-state situations, in modeling the heat transport along nanowires or thin layers, some of these effects may be incorporated into a size-dependent effective thermal conductivity λ_{eff} [2, 43], and a Fourier law (FL)-type equation may still be used with λ_{eff} as the thermal conductivity, instead of the bulk value λ. However, in fast perturbations, or under strong heat gradients, or in axial geometries an effective thermal conductivity is not enough to overcome the different problems related to the FL, as for instance, the infinite speed of propagation of thermal disturbances, or some genuinely nonlinear effects in steady states [9, 17, 25, 28, 30, 38]. Therefore, in modeling heat conduction, it is necessary to go beyond FL by introducing more general heat-transport equations, and analyze more general geometries than those considered in Chaps. 3 and 4. In Chap. 2 the nonlinear heat-transport equation (2.16) has been introduced. Here we will analyze some consequences of it.

5.1 Flux Limiters and Effective Thermal Conductivity of Short Carbon Nanotubes

Limited values for the speed of propagation of thermal signals directly imply that, for a given energy density, the heat flux cannot reach arbitrarily high values, but it will be bounded by a maximum saturation value, of the order of the energy density times the maximum speed [30]. This can be found, for example, in radiative heat transfer, wherein the maximum heat flux is proportional to the fourth power of the temperature times the speed of light [32, 33, 37]. Similarly, in plasma physics, the maximum attainable heat flux is of the order of $k_B T \sqrt{k_B T / m_e}$, m_e being the electron mass [48].

© Springer International Publishing Switzerland 2016   109
A. Sellitto et al., *Mesoscopic Theories of Heat Transport in Nanosystems*,
SEMA SIMAI Springer Series 6, DOI 10.1007/978-3-319-27206-1_5

The classical FL, predicting infinite speed of propagation for thermal perturbations, is obviously not able to face with these saturation effects, which, instead, may be described by the introduction of nonlinear terms in the evolution equation of the heat flux, as in Eq. (2.16).

In order to check the ability of Eq. (2.16) in describing such saturation effects, let us consider a one-dimensional system in steady-state situations. In this case, the local heat flux $\mathbf{q}$ has only one component along the z axis which, in steady states, is constant everywhere along the system, and Eq. (2.16) reduces to

$$q = -\lambda \left(1 + \xi q^2\right) \nabla_z T. \tag{5.1}$$

This nonlinear generalization of the FL is interesting since it allows to give a deeper physical insight to the parameter ξ therein in terms of an effective nonlinear thermal conductivity, defined as

$$\lambda_{\text{eff,nl}} = \lambda \left(1 + \xi q^2\right). \tag{5.2}$$

From Eq. (5.2) one can firstly observe that ξ has to be negative, since positive values for it do not have any physical meaning. In fact, if $\xi > 0$ then from Eq. (5.2) one may conclude that such a nonlinear thermal conductivity increases for increasing heat flux without bound. This is tantamount to assume that it should be possible to perform a thermodynamic process yielding an infinite value for the heat flux, in contradiction with what previously observed. Furthermore, if $\xi > 0$ the steady states would be unstable, because a random increase of q at a constant temperature difference would lead to a higher thermal conductivity and, therefore, to a still higher value of q.

We note that the same conclusion about the sign of ξ in the effective thermal conductivity (5.2) is derived in the TM theory [7, 22, 51–53] where it is set $\xi q^2 = -M_{\text{tm}}^2$, with M_{tm} given by Eq. (2.28).

Once the sign of the nondimensional parameter ξ has been established on pure physical basis (i.e., $\xi < 0$), according with the second law of thermodynamics prescribing that $\lambda_{\text{eff,nl}}$ defined by Eq. (5.2) cannot become negative, then it is possible to see that Eq. (5.2) describes a reduction in the effective thermal conductivity for increasing values of the heat flux. In particular, for any thermodynamic process, it predicts that the upper bound

$$q_{\max} = 1/\sqrt{\|\xi\|}$$

for the heat flux is asymptotically achieved when the temperature gradient tends to infinity.

In the literature, such a situation is well-known as "flux limiters" [3, 28, 32]. Flux limiters are commonly used, for instance, in radiation hydrodynamics [3, 32] wherein the maximum conceivable radiation energy flow will be $e\bar{a}$, e being the energy density of radiation per unit volume, and $\bar{a}$ the modulus of the speed of light (or the phonon energy density and the modulus of the average phonon speed, respectively, in phonon hydrodynamics). If this interpretation is used, one would set in Eq. (5.2)

$$\xi = -(e\bar{a})^{-2} \tag{5.3}$$

and from Eq. (5.1) the following constitutive equation for q ensues:

$$q = \left(\frac{\sqrt{1 + \Psi^2} - 1}{\Psi} \right) e\bar{a} \tag{5.4}$$

wherein

$$\Psi = 2 \left(\frac{\lambda}{e\bar{a}} \right) \frac{\Delta T}{L} \tag{5.5}$$

with ΔT being the temperature difference applied to the ends of the system, L its longitudinal length, and in deriving Eq. (5.5) we have used the approximation $\nabla_z T = \Delta T / L$

In Chap. 3, we have showed that useful information on the system behavior can be obtained by focusing the attention on the effective thermal conductivity, defined as in Eq. (3.2). Equations (5.4) and (5.5) allow to obtain explicitly the effective thermal conductivity in Eq. (5.2) as

$$\lambda_{\text{eff}} = \frac{qL}{\Delta T} = 2\lambda \left(\frac{\sqrt{1 + \Psi^2} - 1}{\Psi^2} \right) \tag{5.6}$$

which yields the following limiting values:

$$\Psi \to 0 \implies \lambda_{\text{eff}} \to \lambda$$
$$\Psi \to \infty \implies \lambda_{\text{eff}} \to \frac{e\bar{a}L}{\Delta T}.$$

In particular, in the latter situation the maximum heat flux obtained is $q_{\max} = e\bar{a}$.

The formalism of flux limiters, expressed in terms of a nonlinear effective thermal conductivity as in Eq. (5.6), can be applied to short carbon nanotubes, the characteristic size of which is smaller than phonon mean-free path (which is of the order of 700 nm at the room temperature), because in this case the heat transport is practically ballistic, and the heat flux would be expected to be of the order of the phonon energy density times the phonon velocity [52, 53].

To use concrete experimental data, we consider that when the temperature is very low, the specific heat at constant volume and the thermal conductivity of the carbon nanotube are, respectively [6, 23]

$$c_v = \frac{3.292}{\pi} \left(\frac{k_B^2 T}{\hbar A \bar{a}} \right) \tag{5.7a}$$

$$\lambda = c_v \bar{a} \ell_p \tag{5.7b}$$

with $\hbar$ as the reduced Planck constant, and A as the cross-sectional area of the cylindrical wall of carbon atoms forming the nanotube. Moreover, from Eq. (5.7a) the phonon energy per unit volume is

$$u = \int c_v(T)\, dT = \frac{3.292}{2\pi}\left(\frac{k_B^2 T^2}{\hbar A \bar{a}}\right) = \frac{c_v T}{2} \tag{5.8}$$

which yields that the nondimensional parameter Ψ in Eq. (5.5) is

$$\Psi = 4\,\mathrm{Kn}\,\frac{\Delta T}{T} \tag{5.9}$$

wherein we set $\mathrm{Kn} = \ell_p/L$. In this way, the introduction of Eq. (5.9) in Eqs. (5.4) and (5.6) allows to obtain the behaviors of the heat flux and of the effective thermal conductivity, respectively, as a function of the longitudinal length L of the nanotube.

These behaviors are shown in Figs. 5.1 and 5.2, respectively, for a carbon nanotube when the average temperature (taken as a reference level for the sake of computation) is $T = 30\,\mathrm{K}$ ($\ell_p \sim 0.5 - 1.5\,\mu\mathrm{m}$, and $\lambda \sim 60 - 180\,\mathrm{Wm^{-1}\,K^{-1}}$ [23]), and when the diameter of the wire is equal to $1.4\,\mathrm{nm}$ and its transversal section has an area equal to $A = 2.5\,\mathrm{nm^2}$. In each figure, three different values for the difference

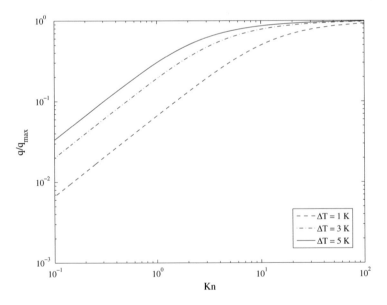

Fig. 5.1 Behavior of the ratio between the local heat flux q, predicted by Eq. (5.4), and the maximum attainable local heat flux $q_{\max} = e\bar{a}$, as a function of $\mathrm{Kn} = \ell_p/L$ in carbon nanotube when the average temperature is approximatively equal to $30\,\mathrm{K}$. For the sake of computation, we assumed $\ell_p = 10^{-6}\,\mathrm{m}$ [23]. The results are shown for three different values of the temperature difference ΔT through the ends of the system, i.e., $\Delta T = 1\,\mathrm{K}$, $\Delta T = 3\,\mathrm{K}$ and $\Delta T = 5\,\mathrm{K}$. In figure both the x-axis, and the y-axis are in a logarithmic scale

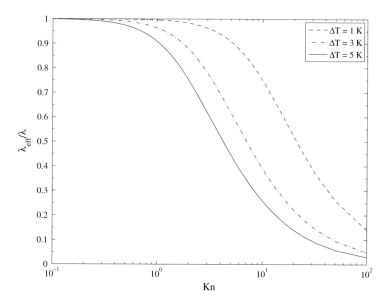

Fig. 5.2 Behavior of the ratio between the effective thermal conductivity λ_{eff}, predicted by Eq. (5.6), and the bulk thermal conductivity λ, as a function of $Kn = \ell_p/L$ in carbon nanotube when the average temperature is approximatively equal to 30 K. For the sake of computation, we assumed $\ell_p = 10^{-6}$ m [23]. The results are shown for three different values of the temperature difference ΔT through the ends of the system, i.e., $\Delta T = 1$ K, $\Delta T = 3$ K and $\Delta T = 5$ K. In figure the x-axis is in a logarithmic scale

temperature through the ends of the nanotube have been chosen, i.e., $\Delta T = 1$ K, $\Delta T = 3$ K, and $\Delta T = 5$ K.

In more details, Fig. 5.1 plots the behavior of the local heat flux q referred to its maximum attainable value (that is, $q_{\max} = e\overline{a}$) as a function of the Knudsen number Kn. As it is possible to observe, the smaller L (that is, the bigger Kn), the higher q flowing through the nanotube at a given ΔT. This is logical because short values of L implies high values for the temperature gradient $\Delta T/L$. Note that, once the reference temperature of the system has been chosen, then the mean-free path of heat carriers is fixed, and therefore changes in the Knudsen number are only due to changes in L. Moreover, as it was expected, the bigger ΔT, the bigger q, systematically. Note, however, that for high Kn, q becomes independent of L and ΔT.

In Fig. 5.2, the ratio between the effective thermal conductivity λ_{eff} and the bulk thermal conductivity λ, is plotted as a function of $Kn = \ell_p/L$. It turns out that the smaller L (that is, the bigger Kn), the smaller λ_{eff} of the nanotube. In particular, when the longitudinal length L switches from the micrometer to the nanometer length scale (i.e., when Kn spans in the range $\left[10^{-1}; 10^2\right]$), the effective thermal conductivity predicted by Eq. (5.6) reduces of one order of magnitude. The asymptotic behavior of the effective thermal conductivity is reached both whenever Kn gets small values, and when it reaches high values. Moreover, the smaller ΔT, the bigger λ_{eff}. This is expected because, as it was shown in Fig. 5.1, the heat flux increases for increasing temperature differences, and therefore, according to Eq. (5.6), the effective thermal conductivity has to decrease.

5.2   Thermal Rectification in Tronco-Conical Nanowires

The possibility of controlling phonon heat transport by means of suitable nanoengineering of materials has opened the possibility to achieve thermal diodes, thermal transistors, thermal logic gates and thermal memories. This is a fast developing area in heat transport, dubbed as *phononics*, in analogy with electronics [35]. From the conceptual point of view, it is a very challenging topic which provides a stimulus for advanced research.

Rectifiers, that is, devices which are able to transport current efficiently in one direction of the applied bias, while blocking it completely (or significantly) in the reverse direction, have a wide interest in electron transfer. Electrical rectifiers may take several different forms, including vacuum tube diodes, mercury-arc valves, copper and selenium oxide rectifiers, semiconductor diodes, silicon-controlled rectifiers and other silicon-based semiconductor switches. Their invention was the marker point of the emergence of modern electronics.

It is apparent that counterpart devices for heat conduction, i.e., thermal rectifiers [12–14, 34, 49, 55], would have interesting implications because of the impact they could have on electronics cooling research, as well as solid-state energy conversion with the ability to control transport. In recent years, some theoretical proposals for thermal rectifiers have been put forward, but these usually require complex coupling between individual atoms and substrates that are difficult to achieve experimentally [42]. In principle, if the thermal conductivity is only a function of temperature, there will not be heat rectification. This possibility arises if thermal conductivity depends on the radius of the component of the system, besides on temperature, and these variables change along the system. For instance, it is well-known that if a system is composed of two parts in series, A and B, and if the thermal conductivity of A increases with increasing T whereas that of B decreases for the same values of temperature, the global thermal conductivity will be higher when A is on the hot side and B at the cold side, than in the opposite case. Thus, the heat flux from A to B will be higher than that from B to A.

In the presence of explicit nonlinear effects, rectification effects may be enhanced. In the present section we will consider such a case. However, as noted by Peierls [41], heat transport in one-dimensional systems can be anomalous, and the breakdown of FL in those systems may be coupled with extraordinary nonlinear thermal effects, including rectification. Nanotubes are nearly one-dimensional systems, and thus they are ideal structures for exploring thermal rectification effects.

In order to investigate the role of nonlinear terms in thermal rectification, consider a tronco-conical nanowire, namely, a nanotube which does not have a

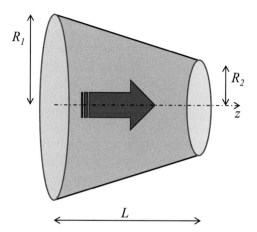

Fig. 5.3 Sketch of a tronco-conical nanowire. The radius of the *left-hand section* is R_1, and that of the *right-hand section* is R_2. For the sake of illustration, we suppose $R_1 > R_2$. The *dashed-dotted line* represents the longitudinal z-axis of the system, the length of which is L. The *red arrow* stands for the local heat flux **q**. Moreover, we assumed that the R_1-section is the incoming section, and the R_2-section is the outgoing one, that is, we assumed $Q_0 > 0$. This means that T_1 (i.e., the temperature evaluated in the *left-hand side section*) is higher than T_2 (i.e., the temperature evaluated in the *right-hand side section*). The opposite situation may also occur. In that case $T_2 > T_1$, and the heat flow from the *right-hand side* to the *left-hand side* (i.e., $Q_0 < 0$)

constant transversal area since the radius R of the cross section changes along the longitudinal z-axis (see Fig. 5.3 for a qualitative sketch of the geometry). Of course, an analogous approach could be made in two-dimensional systems, namely, in ribbons or plane thin layers with inhomogeneous width.

Tronco-conical devices have been studied in Ref.[55] as a possible heat rectifier, with λ and c_v varying on temperature and size.[1] Here, instead, we are mainly interested in analyzing the consequences on thermal rectification arising from genuinely nonlinear terms appearing in the evolution equation of the heat flux. In particular, for the sake of illustration, we suppose that the tronco-conical device is characterized by a transversal section whose area progressively changes along the longitudinal axis because its radius has a linear profile given by [45]

$$R(z) = R_1 - \zeta z, \tag{5.10}$$

where $\zeta = (R_1 - R_2)/L$, with R_1 as the radius of the incoming section, R_2 as the radius of the outgoing section, and L as the longitudinal length of the nanowire. If one ignores, for the sake of simplicity, the radial component of the heat flux, in steady states the local heat flux is such that

$$q\pi R^2(z) = Q_0 \tag{5.11}$$

---

[1] Of course, the latter is only true in nanodevices.

with Q_0 being the total heat flux across any transversal section. Consequently, in this case Eq. (2.16) reduces to

$$\left[1 + \left(\mu + \mu'\right) \nabla_z q\right] q = -\lambda \left(1 + \xi q^2\right) \nabla_z T + \ell_p^2 \nabla_z^2 q. \tag{5.12}$$

The role of the coefficients μ and μ' in Eq. (5.12) can be analyzed by observing that they have the dimensions of a reference length divided by a reference heat flux [45], and then they are expected to be of the order of $\ell_p/(e\bar{a})$. Recalling the meaning of the coefficients ξ given in Sect. 5.1, in Eq. (5.12) the relative importance of the physical effects due to the term $\lambda \xi q^2 \nabla_z T$ with respect to those due to the term $(\mu + \mu') (\nabla_z q) q$ can be estimated by the nondimensional ratio

$$\frac{\lambda \xi q^2 \nabla_z T}{(\mu + \mu') (\nabla_z q) q} \propto \left(\frac{q^2}{\|\nabla_z q\|}\right)\left(\frac{1}{e\bar{a}\ell_p}\right) = \left(\frac{\|q\|}{e\bar{a}}\right)\left(\frac{L}{\ell_p}\right) \approx \frac{\|q\|}{\|q_{max}\|}.$$

Thus, whenever the local heat flux gets values much smaller than the maximum value of the local heat flux, then Eq. (5.12) reduces to

$$\left[1 + \left(\mu + \mu'\right) \nabla_z q\right] q = -\lambda \nabla_z T + \ell_p^2 \nabla_z^2 q \tag{5.13}$$

and the coupling of Eq. (5.13) with Eq. (5.11) turns out [45]

$$\lambda \frac{dT}{dz} = -\frac{Q_0}{\pi R^2} \left\{1 + \frac{2\zeta}{R^2}\left[\left(\mu + \mu'\right)\frac{Q_0}{\pi R} - 3\zeta\ell_p^2\right]\right\} \tag{5.14}$$

which draws out interesting consequences leading to thermal rectification.

When the temperature difference $T_2 - T_1 < 0$ with respect to the positive direction of z-axis in Fig. 5.3, then $Q_0 > 0$, i.e., the heat flows from the left to the right. In this case, from Eq. (5.14) one has

$$Q_0^{(lr)} = \frac{B_1}{2B_2}\left[\sqrt{1 + 4\lambda\left(\frac{B_2}{B_1^2}\right)\left(\frac{\Delta T}{L}\right)} - 1\right] \tag{5.15}$$

where $Q_0^{(lr)}$ denotes the heat flowing from the left-hand section to the right-hand section and

$$B_1 = \frac{1}{\pi R_m^2}\left(1 - 6\zeta^2 \frac{\ell_p^2}{R_m^2}\right) \tag{5.16a}$$

$$B_2 = \frac{2\zeta}{\pi R_m^5}\left(\mu + \mu'\right) \tag{5.16b}$$

where $R_m = (R_1 + R_2)/2$, and $\Delta T = |T_2 - T_1|$.

In the opposite situation, namely, when the temperature difference $T_2 - T_1 > 0$, then $Q_0 < 0$ (with respect to the positive direction of z in Fig. 5.3) and the heat flows from the right to the left. In this case by means of Eq. (5.14) the heat flowing from the right to the left is

$$Q_0^{(\mathrm{rl})} = \frac{B_1}{2B_2} \left[\sqrt{1 - 4\lambda \left(\frac{B_2}{B_1^2}\right)\left(\frac{\Delta T}{L}\right)} - 1 \right]. \qquad (5.17)$$

Up to the lowest-order approximation in $\Delta T/L$, from Eqs. (5.15) and (5.17) the following thermal-rectification ratio, defined as the ratio of the values of the heat flux in the reverse (right-left) and the direct (left-right) direction, is recovered

$$\left|\frac{Q_0^{(\mathrm{rl})}}{Q_0^{(\mathrm{lr})}}\right| = \frac{1 - \sqrt{1 - 4\lambda \left(\frac{B_2}{B_1^2}\right)\left(\frac{\Delta T}{L}\right)}}{\sqrt{1 + 4\lambda \left(\frac{B_2}{B_1^2}\right)\left(\frac{\Delta T}{L}\right)} - 1} \approx \frac{1 + 2\lambda \left(\frac{B_2}{B_1^2}\right)\left(\frac{\Delta T}{L}\right)}{1 - 2\lambda \left(\frac{B_2}{B_1^2}\right)\left(\frac{\Delta T}{L}\right)}$$

$$\approx 1 + 4\lambda \left(\frac{B_2}{B_1^2}\right)\left(\frac{\Delta T}{L}\right). \qquad (5.18)$$

Note that if $(\mu + \mu') = 0$, then $B_2 = 0$ and the ratio above tends to 1, namely, no thermal rectification is obtained.

Although the thermal-rectification phenomenon may be also described if the temperature and size dependence of λ and c_v is taken into account [55], the results above show the role played by genuinely nonlinear effects, which are strictly related to the coefficients μ and μ', in the evolution equation of $\mathbf{q}$. In practical applications, it would be interesting to relate those coefficients with well-known physical quantities. The rigorous way to do this is by direct measurements, or from experimental simulations. Indeed, a first rough, but simple, estimation of those coefficients could be obtained by a comparison with other theoretical models. In particular, a comparison with the model proposed in Refs. [10, 11] turns out that

$$\mu = -\frac{2\tau_p}{c_v T} \qquad (5.19)$$

whereas the comparison with the model proposed in the Thermomass (TM) theory [7, 22, 51] yields

$$\mu' = \frac{\tau_{\mathrm{tm}}}{c_v T}. \qquad (5.20)$$

5.3 Axial Heat Transport in Thin Layers and Graphene Sheets

Systems with large values of the mean-free path for heat carriers are expected to show some different features in heat transport than those obeying the classical FL [45, 46]. Up to the first order, a long mean-free path implies a high thermal conductivity, consistent with FL. However, in a more general setting, it also implies long collision times and, therefore, memory effects, which are not included in the FL, but which will be relevant in high-frequency perturbations and thermal-wave propagation, as well as effects of second or higher-order in the mean-free path. Steady-state situations allow to focus the attention only on nonlocal effects beyond the FL, since in this case memory effects are not relevant. Nonlocal effects will manifest themselves if heat-flux (or temperature) inhomogeneities are sufficiently steep. One-dimensional temperature perturbations have been much studied in such systems, but two-dimensional perturbations may provide some new perspectives on nonlocal effects, related to the influence of the mean-free path on the steady-state temperature profile, which are not directly accessible in one-dimensional problems. These effects could be exhibited by radial heat transport in silicon thin layers, or in graphene sheets, which are systems with relatively large mean-free paths (for example, at the room temperature, the phonon mean-free path is of the order of 10^{-7} m [4, 5, 19, 20, 29]). This kind of situations are of interest for some thermodynamic technologies, or in the refrigeration of nanodevices.

As an example, here we consider the radial heat transport from a point source in a flat system, and explore the possible consequences of nonlocal and nonlinear effects. In more detail, we suppose that a cylindrical nanodevice (the base-radius of which is R_0) acts as a steady heat source (the temperature of which is held at the constant value T_0) and is connected to a surrounding thin layer, which removes the heat from that nanodevice (see Fig. 5.4 for a qualitative sketch).

We assume the steady-state situation, in which the heat removed is equal to the heat dissipated, thus allowing the temperature of the system to remain constant. In this situation the layer is supplied with a constant amount of heat Q_0 per unit time which propagates radially away from that source, and we look for the influence of nonlocal effects on the radial temperature profile.

For the sake of simplicity, it is possible to assume that the layer is isotropic, namely, that the heat propagates in the same way along all radial directions in the external layer, and $\xi = 0$ in Eq. (2.16).

Due to the axial symmetry, in steady-state situations the heat flux in the surrounding layer has only one component. To obtain the radial dependence of q, one may observe that in the absence of transversal exchanges, the total heat flux flowing across each concentric circular area is always the same. Thus, once two different concentric circular areas of radial distance from the source equal to r and

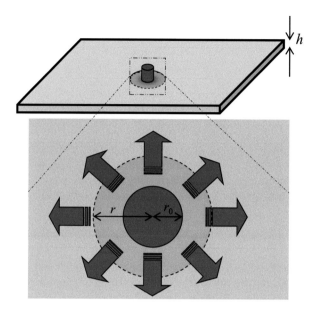

Fig. 5.4 Cylindrical nanodevice connected to a graphene layer. The internal device (*red*) is characterized by a radius r_0. The surrounding thin layer (*grey*) has, instead, a radial characteristic size which is much larger than r_0, and a thickness h. The surrounding layer removes heat from the core device, which acts so as a steady heat source. The heat per unit time produced by the inner hot component is Q_0. Abroad the hot spot, the local heat flux **q** propagates radially away from the source. This is shown in the figure zoom, wherein a further circular zone of radius r is represented

$r + dr$ have been chosen, then the radial profile of the local heat flux is such that

$$2\pi r q (r) = 2\pi \, [r + dr] \, q \, (r + dr) \approx 2\pi \, [r + dr] \left[q \, (r) + \frac{dq}{dr} dr \right]$$

$$\Rightarrow r\frac{dq}{dr} + q \, (r) = 0$$

when a first-order approximation in dr is used. The relation above may be easily integrated in order to have

$$q \, (r) = \frac{\Gamma}{r} \tag{5.21}$$

with $\Gamma = Q_0 / (2\pi h)$, h being the thickness of the layer (see Fig. 5.4). Under previous assumptions, the introduction of Eq. (5.21) in Eq. (5.13) leads to

$$\lambda \frac{dT}{dr} = \frac{\Gamma}{r} \left[\frac{\ell_p^2 + \Gamma \, (\mu + \mu')}{r^2} - 1 \right] \tag{5.22}$$

the integration of which allows to obtain the radial temperature profile. From
Eq. (5.22), in particular, one has [45]

$$
\begin{cases}
r < \sqrt{\ell_p^2 + \Gamma\,(\mu + \mu')} \Rightarrow \dfrac{dT}{dr} > 0 \\[2mm]
r > \sqrt{\ell_p^2 + \Gamma\,(\mu + \mu')} \Rightarrow \dfrac{dT}{dr} < 0
\end{cases}
\tag{5.23}
$$

which means that in the layer which surrounds the hot component there may be both
radial distances for which the temperature profile is increasing, and radial distances
for which the temperature profile is decreasing. From the physical point of view,
previous results point out that the temperature profile shows a hump reached at

$$
r = \sqrt{\ell_p^2 + \Gamma\,(\mu + \mu')}.
$$

If the identifications (5.19) and (5.20) for the coefficients μ and μ' hold, then the
difference $2\tau_p - \tau_{tm}$ has the relevant role of shortening or enlarging the width of the
region wherein the temperature hump appears.

Note that recently the role of hydrodynamic phonon transport in graphene has
been the subject of increasing attention, as it is seen that it occurs in graphene at
higher temperatures than in the bulk materials, and it makes a relevant contribution
to heat transport [8, 19, 31].

5.3.1 Nonlocal Heat Transport and Steady-State Temperature Profile in Thin Layers

In integrated circuits the number of transistors that can be placed on a single chip
has increased exponentially, since the developments in nanotechnology allow the
transistor to shrink in their sizes. With more transistors integrated on a chip, running
at faster clock rate, designers have been able to improve the system performance
effectively over the past few decades. Unfortunately, the design complexity, as well
as the increase in the computational rate, much enhances the rate of heat per unit
volume which has to be dissipated, thus being a limiting factor in current computer
miniaturization. However, the incorporation of silicon or graphene into chip design
could yield devices that are faster, less noisy, and run cooler.

Since the exact forms of the relaxation times is a matter of great discussions, and
all the different estimations of them one can find in literature are related to suitable
adjustable coefficients [15, 16, 54], in order to deal with the simplest situation, just
for the sake of illustration here we investigate the consequences of the radial heat
propagation (5.21) in the absence of the nonlinear terms in $(\mu + \mu')\nabla_z q$. In this

case Eq. (5.22) reduces to

$$\lambda \frac{dT}{dr} = \frac{\Gamma}{r}\left[\left(\frac{\ell_p}{r}\right)^2 - 1\right].$$ (5.24)

5.3.1.1   Silicon Thin Layers

Silicon thin layers are widely applied in the nanoelectronics industry for the fabrication of transistors, solid-state lasers, sensors and actuators. Since the thermal-transport properties affect the performance and reliability of these devices, a better understanding of their thermal behavior is useful for practical applications. Indeed, in two-dimensional situations, nonlocal terms affect the heat transport both in the radial r direction (i.e., along the plane of the layer), and in the transversal y direction (i.e., in the perpendicular direction to the layer, as it is shown in Fig. 5.5).

The consequences of those terms along the latter direction may be accounted, for example, by means of a phonon-hydrodynamic approach, as shown in Chap. 3. In particular, whenever $Kn_p > 1$ (defined in this case as $Kn_p = \ell_p/h$), in Eq. (5.24) one should replace the thermal conductivity λ with its effective value which, in a first-order approximation, can be approximated by the expression in Eq. (3.32).

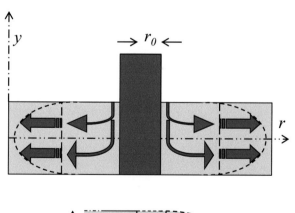
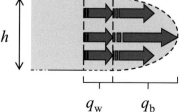

Fig. 5.5 Theoretical heat profile in a thin layer with thickness h: frontal view. The hot component is sketched as a *red rectangle*. To account for the finite value of h, by means of phonon hydrodynamics we assume that the longitudinal heat flux has a bulk contribution q_b (with a parabolic profile) and a wall contribution q_w [2] (see the sketch in bottom side of the figure). The layer's vertical size is emphasized with respect the longitudinal ones just for the sake of illustration. In practical applications, in fact, the thickness of the layer is much smaller than its width

Then, the integration of the modified version of Eq. (5.24) gets the following radial behavior of the temperature profile [46]:

$$
\begin{cases}
T(r) = T_0 & \forall r \leq r_0 \\
T(r) = T_0 + \dfrac{\Gamma}{2\lambda_{\text{eff}}} \left[\dfrac{\ell_p^2}{r_0^2} \left(1 - \dfrac{r_0^2}{r^2} \right) + 2 \ln \left(\dfrac{r_0}{r} \right) \right] & \forall r > r_0
\end{cases}
\tag{5.25}
$$

which points out that in a range of radial distances, the order of which is a few times the mean-free path ℓ_p, the radial behavior of the temperature $T(r)$ is strongly modified by the nonlocal terms, since in their absence the predicted temperature profile would be

$$
\begin{cases}
T(r) = T_0, & \forall r \leq r_0 \\
T(r) = T_0 + \dfrac{\Gamma}{\lambda_{\text{eff}}} \ln \left(\dfrac{r_0}{r} \right), & \forall r > r_0.
\end{cases}
\tag{5.26}
$$

Note that the behavior predicted by Eqs. (5.25) is functionally different from that predicted by Eqs. (5.26), where the nonlocal term yields a further contribution, proportional to r^{-2}, which is lacking in Eq. (5.25) and which can not be described by means of an effective thermal conductivity.

The different behaviors may be also clearly seen in Fig. 5.6, wherein the consequences of nonlocal effects on the radial behavior of the temperature have

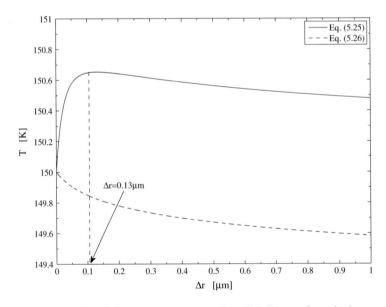

Fig. 5.6 Radial behavior of the temperature versus the radial distance from the hot spot (i.e., $\Delta r = r - r_0$) in a silicon thin layer ($\lambda = 336.27\,\mathrm{W\,m^{-1}\,K^{-1}}$, $\ell_p = 3.08 \cdot 10^{-7}$ m, and $C = 0.36$) with $h = 50 \cdot 10^{-9}$ m heated at the center. Comparison between the predictions of Eqs. (5.25) and (5.26). The heat source is characterized by $Q_0 = 10^{-6}$ W, $T_0 = 150$ K, and $r_0 = 50 \cdot 10^{-9}$ m

been shown for a silicon thin layer with $h = 50 \cdot 10^{-9}$ m. In the computation we have supposed that the layer is heated with $Q_0 = 10^{-6}$ W, the hot-spot temperature is $T_0 = 150$ K, and its radius is $r_0 = 50 \cdot 10^{-9}$ m. We have chosen such a value for r_0 since we are interested in a small value of the radius, smaller than the phonon mean-free path, and we have taken as an indicative size that of the layer thickness. However, the effect of the radius will be further considered below, in Fig. 5.7. The wall-accommodation parameter at 150 K is $C = 0.36$ and refer to a smooth wall [44], so that the effective thermal conductivity is $\lambda_{eff} = 23.2$ Wm^{-1} K^{-1}.

As it is possible to observe, the temperature behavior arising from Eq. (5.25) (solid lines in Fig. 5.6) shows a hump (with a maximum reached when the radial distance from the hot spot is $\Delta r = 1.3 \cdot 10^{-7}$ m), namely, when the radial distance from the hot spot is smaller than the phonon mean-free path, then the temperature increases with the radius. This does not appear in the temperature behavior given by Eq. (5.26) (dashed lines in Fig. 5.6) which, instead, is always decreasing.

From Fig. 5.6 we infer that the deviation ΔT of the temperature predicted by Eq. (5.25) from that predicted by Eq. (5.26), i.e.,

$$\Delta T (r) = \frac{\Gamma}{2\lambda_{eff}} \left(1 - \frac{r_0^2}{r^2} \right) \frac{\ell_p^2}{r_0^2} \tag{5.27}$$

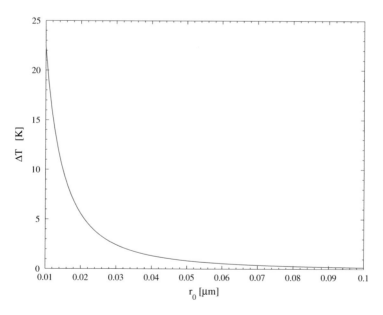

Fig. 5.7 Behavior of the temperature difference ΔT, arising from Eq. (5.27), as a function of r_0 when the temperature of the hot spot is $T_0 = 150$ K and $Q_0 = 10^{-6}$ W, in a silicon thin layer with $h = 50 \cdot 10^{-9}$ m. The temperature difference is evaluated at a distance from the hot spot which is equal to the phonon mean-free path, namely, $\Delta r \equiv \ell_p = 1.81 \cdot 10^{-7}$ m

is small (around one degree at 150 K), but it could be measured in principle. However, since ΔT is inversely proportional to λ_{eff}, it would still be higher in the case of rough-walled thin layers [43, 44], where the effective thermal conductivity is an order of magnitude smaller than that in smooth-walled thin layers.

It is also possible to note that a way to enhance that difference would be either to increase the amount of heat per unit time Q_0, or to reduce the heated-spot radius r_0. In Fig. 5.7, the variation of ΔT in terms of r_0 is examined, at a distance $r = \ell_p$ from the heated spot, for fixed values of h, Q_0 and T_0 (they are the same used to obtain the results in Fig. 5.6). Figure 5.7 points out that the smaller r_0, the greater ΔT.

5.3.1.2   Graphene Sheets

Graphene is a flat monolayer of carbon atoms tightly packed into a two-dimensional honeycomb lattice [5, 19, 20], which can be wrapped up into zero-dimensional fullerenes, rolled into one-dimensional nanotubes, or stacked into three-dimensional graphite. This means that in the case of graphene one could practically speak about a sheet, or a very thin layer, the thickness of which is $h \approx 3 \cdot 10^{-10}$ m, namely, of the order of the atomic diameter of carbon. At room temperature, the measured thermal conductivity of graphene λ_{eff} is found to span the range $[3080; 5150]\,\mathrm{Wm}^{-1}\,\mathrm{K}^{-1}$, and the phonon mean-free path is of the order of $\ell_p = 7.75 \cdot 10^{-7}$ m [4, 20, 39, 40].

The thermal properties of graphene have been reviewed in detail by Balandin [4], with further information about analogous properties in nanostructured carbon materials. It is pointed out how graphene (or other materials able to conduct heat well) could be essential for the refrigeration of small devices, as in the next generation of highly miniaturized integrated circuits, or optoelectronic and photonic devices, by removing the heat dissipated during their functioning at very high frequencies.

Figure 5.8 plots the radial temperature profile arising from the theoretical model in Eq. (5.24) in the graphene sheet, as a function of the radial distance r from the hot spot at $T_0 = 300$ K. For the sake of computation, in obtaining the behaviors in that figure it has been supposed that the heat per unit time produced by the inner hot component is $Q_0 = 10^{-8}$ W.

In order to show the influence of the characteristic size of the heat source, in Fig. 5.8 two different values for the radius r_0 of the internal device have been considered, i.e., $r_0 = 15 \cdot 10^{-9}$ m and $r_0 = 50 \cdot 10^{-9}$ m. In particular, Fig. 5.8 points out that the smaller r_0, the bigger the reached temperature abroad the hot zone. Moreover, it can be seen that the effects of nonlocal terms are felt at distances which are several times larger than the mean-free path of heat carriers. The range of influence of nonlocal terms is evidenced by supposing that the radial distance from the center of the hot spot extends till 10^{-4} m, whereas in practical applications the largest size in nanosystems should not exceed 10^{-7} m.

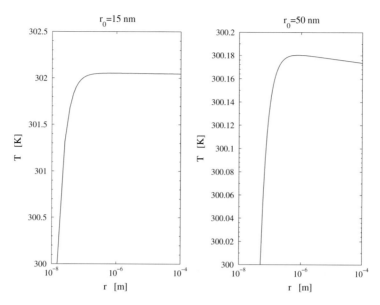

Fig. 5.8 Radial behavior of the temperature in a graphene sheet with $h_g = 3.35 \cdot 10^{-10}$ m heated at the center. The heat source is characterized by $Q_0 = 10^{-8}$ W, and $T_0 = 300$ K ($\lambda = 3080$ Wm^{-1} K^{-1}, and $\ell_p = 7.75 \cdot 10^{-7}$ m [5, 20]). Two different values of the hot spot radius have been considered, that is, $r_0 = 15 \cdot 10^{-9}$ m and $r_0 = 50 \cdot 10^{-9}$ m. Two different length scales for the temperature have been used, in order to show clearly both results. The x-axis is in a logarithmic scale

5.3.2 Thermodynamic Aspects: Second Law at the Mean-Free Path Scale

By means of Eq. (5.25) it has been observed that the radial dependence of temperature profile has a surprising behavior in the annular region between r_0 and ℓ_p, since in it the temperature increases with the radius, instead of being decreasing, as it would have been expected on classical grounds and predicted by the classical FL. The decrease is found, instead, whenever $r > \ell_p$. Indeed, the reality of that result should be experimentally checked; here we discuss its compatibility with the classical statement of second law of thermodynamics.

From the microscopic point of view, a possible interpretation of this phenomenon is the following one. In the central region, heat is intense and the particles have a relatively high energy in the radial direction. They ballistically follow such direction until they have their first collisions. Then, their radial energy is redistributed among other particles, and other directions. This makes that the temperature of such regions increases.

From the thermodynamic point of view, however, the situation is challenging. If the classical local-equilibrium formulation of the second law of thermodynamics (3.53) holds, then the use of Eq. (5.21) for $q(r)$ and of Eq. (5.25) for dT/dr leads to the conclusion that the entropy production in the local-equilibrium approximation would be

$$\sigma_{\text{le}}^{s} = \frac{1}{\lambda T^2} \left(\frac{\Gamma}{r} \right)^2 \left[1 - \left(\frac{\ell_p}{r} \right)^2 \right]$$

(5.28)

which is negative for $r < \ell_p$, i.e., in the region where the temperature increases with the radius. Thus, the temperature hump should be forbidden.

A deeper insight, instead, can be obtained by analyzing the compatibility with second law of thermodynamics in the framework of EIT [28, 30, 36]. In this theory, the extended entropy may be represented by a general constitutive equation of the type $s_{\text{eit}} = s_{\text{eit}}(u, \mathbf{q}, \mathbf{Q})$, which becomes $s_{\text{eit}} = s_{\text{eit}}(u, \mathbf{q}, \nabla \mathbf{q})$ whenever the relaxation time of $\mathbf{Q}$ is sufficiently smaller than that of $\mathbf{q}$, in such a way that the flux of heat flux reduces to $\mathbf{Q} = \ell_p^2 \nabla \mathbf{q}$ [26, 43, 47], and the influence of nonlocal terms is more explicit. In this case the entropy production is given by Eq. (3.54), and its introduction in the coupling of Eqs. (5.21) and (5.25), instead, leads to

$$\sigma_{\text{eit}}^{s} = \frac{1}{\lambda T^2} \left(\frac{\Gamma}{r} \right)^2 \left[1 + \left(\frac{\ell_p}{r} \right)^2 \right]$$

(5.29)

which is everywhere positive. Thus, if the second law is expressed as the requirement of positive right hand side of Eq. (5.29), then the temperature hump is compatible with the second law of thermodynamics.

From this point of view, the hump in the temperature profile observed between r_0 and ℓ_p would have much interest from a basic thermodynamic perspective, because it would explicitly show both the influence of nonlocal effects in heat transport, and the need of generalizing the formulation of the second law. It seems worth observing that a similar hump, in traveling front solution of reaction-diffusion equation with a nonlocal term has been obtained, on theoretically grounds, as a consequence of nonlocalities in the equation for the diffusion flux, but without a thermodynamic analysis [21].

Although Eq. (5.29) ensures the physical consistency of Eq. (5.24), it does not give any information either about the hump, or about the amplitude of the region wherein the heat flows from the colder points to the warmer ones. This means that the analysis of the sign of the entropy production is not able to cut off unreal phenomena, as for example the case of unbounded increasing temperature for increasing radial distance. More detailed conclusions may be obtained, instead, if one focuses the attention on the entropy flux $\mathbf{J}^s$. As we observed in Chap. 1, in EIT [28, 30] it reads

$$\mathbf{J}^s = \frac{\mathbf{q}}{T} + \frac{\ell_p^2}{\lambda T^2} \nabla \mathbf{q}^{\mathrm{T}} \cdot \mathbf{q}$$

(5.30)

and reduces to

$$J^s(r) = \left(\frac{1}{T} - \frac{\ell_p^2}{\lambda T^2} \frac{\Gamma}{r^2} \right) q(r) \qquad (5.31)$$

in the case of radial heat propagation [47]. Restricting the analysis to an annular region between two generic radial distances r_1 and r_2, such that $r_1 < r_2$, then in steady states the entropy balance (1.9) implies

$$2\pi r_1 J^s(r_1) < 2\pi r_2 J^s(r_2) \qquad (5.32)$$

wherein $J^s(r_1)$ is the incoming entropy flux, and $J^s(r_2)$ is the outgoing one. The coupling of Eqs. (5.31) and (5.32) firstly gets

$$2\pi r_1 q(r_1) \left(\frac{1}{T_1} - \frac{\ell^2}{\lambda T_1^2} \frac{\Gamma}{r_1^2} \right) < 2\pi r_2 q(r_2) \left(\frac{1}{T_2} - \frac{\ell^2}{\lambda T_2^2} \frac{\Gamma}{r_2^2} \right). \qquad (5.33)$$

Then, owing to the conservation of the total heat flux across each circular zone, the previous relation implies

$$\frac{1}{T_1} \left(1 - \frac{\ell^2}{\lambda T_1} \frac{\Gamma}{r_1^2} \right) < \frac{1}{T_2} \left(1 - \frac{\ell^2}{\lambda T_2} \frac{\Gamma}{r_2^2} \right). \qquad (5.34)$$

It is easy to see from the relation above that whenever $\lambda T > \ell_p^2 \Gamma / r^2$ (or in the limit case of $\Gamma \ell_p^2 / (\lambda Tr^2) \to 0$), then inequality (5.34) is not violated if $T_1 > T_2$. On the other hand, the condition $\lambda T > \ell_p^2 \Gamma / r^2$ is fulfilled, for example, when $\ell_p < r_1, r_2$. Therefore, we may conclude that for radial distances from the center of the hot point larger than the mean-free path, the larger the radial distance, the smaller the temperature and the possible paradox of unbounded temperature is removed.

Conversely, if $\lambda T < \ell_p^2 \Gamma / r^2$ (this is the case, for example, when $r_1, r_2 < \ell_p$), inequality (5.34) changes in

$$\frac{1}{T_1} \left(\frac{\ell_p^2}{\lambda T_1} \frac{\Gamma}{r_1^2} - 1 \right) > \frac{1}{T_2} \left(\frac{\ell_p^2}{\lambda T_2} \frac{\Gamma}{r_2^2} - 1 \right) \qquad (5.35)$$

which is not violated for $T_1 < T_2$. Thus, for radial distances from the center of the hot point smaller than the mean-free path, the larger the radial distance, the bigger the temperature, so that the temperature hump arises.

Of course, the general form (5.31) of the entropy flux, applicable to several different situations more general than the concrete situation corresponding to Eq. (5.24), indicates that in some situations - as those in Eq. (5.24) - it is possible to have a heat transfer from lower to higher temperature, at the condition that the inhomogeneity in the heat flux is high enough - namely, Γ reaches high values in

Eq. (5.21) - and the system is of the order of (or smaller than) the mean-free path of heat carriers.

5.4 Stability of the Heat Flow in Nanowires

In Sect. 2.3 we dealt with approximated heat-transport equation by introducing some characteristic numbers. Now, let us illustrate by a simple example the possibility of relating the stability of heat flow with the value of those numbers. To this end, let us suppose $M_q \gg 1$ and $\text{Re}_q \ll \text{Kn}_q^2 / \tau_R^r$. From Eqs. (1.8) and (2.30) we have

$$\nabla \cdot \mathbf{q} = 0 \tag{5.36a}$$

$$\dot{\mathbf{q}} - \mu \nabla \mathbf{q} \cdot \mathbf{q} - \iota \nabla^2 \mathbf{q} + \nabla \pi = \mathbf{0} \tag{5.36b}$$

being $\mu = 2/(Tc_v)$, $\iota = \ell_p^2/\tau_R$, and $\pi = \lambda T/\tau_R$. Prescribe then the following boundary values

$$\mathbf{q}(\mathbf{x}; t) = \mathbf{q}^b(\mathbf{x}; t) \ \forall \mathbf{x} \in \partial \Omega, \ \forall t \geq 0 \tag{5.37}$$

as the initial ones

$$\mathbf{q}(\mathbf{x}; 0) = \mathbf{q}^0(\mathbf{x}) \ \forall \mathbf{x} \in \Omega \tag{5.38}$$

where $\nabla \cdot \mathbf{q}^0(\mathbf{x}) = 0$, $\forall \mathbf{x} \in \Omega$, being Ω the spatial domain occupied by the nanowire, and $\partial \Omega$ the boundary of the system, respectively. Note that both $\partial \Omega$ and Ω cannot change in time, since we are referring to a rigid body. It is also worth observing that Eq. (5.36a) is not able to furnish now the function u and hence, in principle, it is not possible to determine the function T appearing in the right-hand side of Eq. (5.36b) from its constitutive equation. As a consequence, T must be regarded now as an additional unknown quantity, as the three components of the heat flux. The four unknowns u and $\mathbf{q}$ have to be calculated by solving the system (5.36). The same situation holds in classical hydrodynamics where, in the presence of the constraint of incompressibility, the pressure cannot be calculated through a constitutive equation but becomes an additional unknown quantity.

From the formal point of view, in the next we denote the solutions of the initial boundary value problem (IBVP) in Eqs. (5.36)–(5.38) as

$$\{\mathbf{q}(\mathbf{x}, t, \mathbf{q}^0) ; \pi(\mathbf{x}, t, \mathbf{q}^0)\} \tag{5.39}$$

and we will study their stability when the initial values $\mathbf{q}^0(\mathbf{x})$ are perturbed. To this end, let us consider the further solution

$$\{\mathbf{q}^a; \pi^a\} = \{\mathbf{q}(\mathbf{x}, t, \mathbf{q}^0 + \delta \mathbf{q}^0) ; \pi(\mathbf{x}, t, \mathbf{q}^0 + \delta \mathbf{q}^0)\} \tag{5.40}$$

which still satisfies Eqs. (5.36) and (5.37), but differs from the initial value (5.38) since

$$\mathbf{q}^a\left(\mathbf{x};0\right) = \mathbf{q}^0\left(\mathbf{x}\right) + \delta\mathbf{q}^0\left(\mathbf{x}\right). \tag{5.41}$$

If the solutions (5.39) and (5.40) behave in the same way for increasing time, we face with stable solutions, otherwise we have instable solutions. To be more explicit, we are interested in determining whether the form which the fields in Eq. (5.39) take as $t \to \infty$ is stable with respect to perturbations in the initial conditions, or not. This problem can be analyzed by introducing the following IBVP

$$\nabla \cdot \delta\mathbf{q} = 0 \tag{5.42a}$$

$$\delta\dot{\mathbf{q}} - \mu\left(\nabla\delta\mathbf{q}\cdot\mathbf{q} + \nabla\mathbf{q}\cdot\delta\mathbf{q} + \nabla\delta\mathbf{q}\cdot\delta\mathbf{q}\right) - \iota\nabla^2\delta\mathbf{q} + \nabla\widetilde{p} = \mathbf{0} \tag{5.42b}$$

$$\delta\mathbf{q}\left(\mathbf{x};t\right) = \mathbf{0}, \ \forall\mathbf{x} \in \partial\Omega, \ \forall t \geq 0, \tag{5.42c}$$

$$\delta\mathbf{q}\left(\mathbf{x};0\right) = \delta\mathbf{q}^0\left(\mathbf{x}\right), \ \forall\mathbf{x} \in \Omega, \tag{5.42d}$$

which governs the evolution of the disturbance $\{\delta\mathbf{q};\tilde{p}\} = \{\mathbf{q}^a - \mathbf{q}; \pi^a - \pi\}$. That way, the stability we are looking for coincides with the stability of the problem in Eqs. (5.42) [24]. If we define the average energy of a disturbance as

$$\mathcal{E}\left(t\right) = \frac{1}{\mathcal{M}\left(\Omega\right)}\int_\Omega |\delta\mathbf{q}|^2\,d\Omega \tag{5.43}$$

being $\mathcal{M}\left(\Omega\right)$ the measure of the volume of Ω, we may call the solution of Eqs. (5.42) stable to the initial perturbations if [24]

$$\lim_{t\to\infty}\frac{\mathcal{E}\left(t\right)}{\mathcal{E}\left(0\right)} = 0. \tag{5.44}$$

In order to apply to a very simple situation this stability criterion, let us suppose that the disturbances are characterized by a small amplitude, so that we have $\delta h/h \ll 1$, being h the generic unperturbed quantity. Without loss of generality, this allows us to disregard nonlinear third-order terms in the perturbations, too. Finally, just for the sake of simplicity, suppose also $\mathrm{St}_q \gg 1$. Along with previous observations, in this case we have $\delta\mathbf{q} = \mathbf{0}$, Eq. (5.42b) becomes

$$\mu\left[\delta\left(\nabla\mathbf{q}\cdot\mathbf{q}\right) + \nabla\delta\mathbf{q}\cdot\delta\mathbf{q}\right] = \nabla\widetilde{p} - \iota\nabla^2\delta\mathbf{q} - \mu\nabla\mathbf{q}\delta\mathbf{q} \tag{5.45}$$

which, having present the hypothesis $\delta h/h \ll 1$, up to the second-order in the perturbations, yields

$$|\delta\mathbf{q}|^2 = \frac{1}{\mu^2\,|\nabla\mathbf{q}|^2}\left|\iota\nabla^2\delta\mathbf{q} - \nabla\widetilde{p}\right|^2. \tag{5.46}$$

Therefore, if $M_q \gg 1$ and $Re_q \ll Kn_q^2 / \tau_R^r$, the heat flow remains stable under initial perturbations if

$$\lim_{t \to \infty} \frac{\int_\Omega \left(\frac{Tc_v}{2\tau_R}\right)^2 \left(\frac{\left|\ell_p^2 \nabla^2 \delta \mathbf{q} - \lambda \nabla \delta T\right|}{|\nabla \mathbf{q}|}\right)^2 d\Omega}{\int_\Omega \delta \mathbf{q}^0 d\Omega} = 0 \qquad (5.47)$$

once we have used previous identifications for μ, ι and π.

For finite values of $\mathcal{E}\,(0)$, from Eq. (5.47) we may conclude that, if the gradient of the heat flux does not tend to zero as $t \to \infty$, the stability of $\mathbf{q}$ will be recovered if the perturbations in the initial values are such that $\ell_p^2 \nabla^2 \delta \mathbf{q} \to \lambda \nabla \delta T$ for increasing time. More generally, it is sufficient that the perturbations in the initial values are such that the integral in the numerator of the right-hand side of Eq. (5.47) is infinitesimal.

The physical interpretation of the result above is the following.

For constant internal energy, the gradient of $\mathbf{q}$ can be interpreted as the thermodynamic force which drives the heat flow along a given direction. In Eq. (5.47) the strength of such a force is represented by the term $|\nabla \mathbf{q}|^2$. On the other hand, the strength of the sole initial perturbation appearing in Eq. (5.47) is represented by the quantity $\ell_p^2 \nabla^2 \delta \mathbf{q} - \lambda \nabla \delta T$. For large t, if the initial perturbation has order of magnitude which is smaller than that of the driving force, then it will be not capable of changing appreciably the direction of $\mathbf{q}$, so that the heat flow remains stable.

On the other hand, if for large t the order of magnitude of the initial perturbation gets values high enough to render finite the integral above, i.e., if it is higher than that of the driving force, hence the perturbation can deviate the heat flux from the direction previously determined by the initial conditions, inducing so disordered and unpredictable evolutions of the flow, as, for instance, turbulence and vorticity.

Finally, let us observe that, since the form of the domain is known, the integral above can be estimated [24], obtaining so a threshold for the initial perturbation at which the turbulent regime arises. Such a threshold can be useful, for example, in calculating the risk of thermal damage in MEMS/NEMS design.

References

1. Alvarez, F.X., Jou, D.: Memory and nonlocal effects in heat transports: from diffusive to ballistic regime. Appl. Phys. Lett. **90**, 083109 (3 pp.) (2007)
2. Alvarez, F.X., Jou, D., Sellitto, A.: Phonon hydrodynamics and phonon-boundary scattering in nanosystems. J. Appl. Phys. **105**, 014317 (5 pp.) (2009)
3. Anile, A.M., Pennisi, S., Sammartino, M.: A thermodynamical approach to Eddington factors. J. Math. Phys. **32**, 544–550 (1991)
4. Balandin, A.A.: Thermal properties of graphene and nanostructured carbon materials. Nat. Mater. **10**, 569–581 (2011)

5. Balandin, A.A., Ghosh, S., Baoand, W., Calizo, I., Teweldebrhan, D., Miao, F., Lau, C.-N.: Superior thermal conductivity of single-layer graphene. Nano Lett. **8**, 902–907 (2008)
6. Benedict, L.X., Louie, S.G., Cohen, M.L.: Heat capacity of carbon nanotubes. Solid State Commun. **100**, 177–180 (1996)
7. Cao, B.-Y., Guo, Z.-Y.: Equation of motion of a phonon gas and non-Fourier heat conduction. J. Appl. Phys. **102**, 053503 (6 pp.) (2007)
8. Capellotti, A., Fugallo, G., Paulatto, L., Lazzeri, M., Mauri, F., Marzari, N.: Phonon hydrodynamics in two-dimensional materials. Nat. Commun. **6**, 6400 (7 pp.) (2015)
9. Cimmelli, V.A.: Different thermodynamic theories and different heat conduction laws. J. Non-Equilib. Thermodyn. **34**, 299–333 (2009)
10. Cimmelli, V.A., Sellitto, A., Jou, D.: Nonlocal effects and second sound in a nonequilibrium steady state. Phys. Rev. B **79**, 014303 (13 pp.) (2009)
11. Cimmelli, V.A., Sellitto, A., Jou, D.: Nonequilibrium temperatures, heat waves, and nonlinear heat transport equations. Phys. Rev. B **81**, 054301 (9 pp.) (2010)
12. Criado-Sancho, J.M., Jou, D.: Heat transport in bulk/nanoporous/bulk silicon devices. Phys. Lett. A **377**, 486–490 (2013)
13. Criado-Sancho, J.M., del Castillo, L.F., Casas-Vázquez, J., Jou, D.: Theoretical analysis of thermal rectification in a bulk Si/nanoporous Si device. Phys. Lett. A **19**, 1641–1644 (2012)
14. Criado-Sancho, J.M., Alvarez, F.X., Jou, D.: Thermal rectification in inhomogeneous nanoporous Si devices. J. Appl. Phys. **114**, 053512 (2013)
15. De Tomas, C., Cantarero, A., Lopeandia, A.F., Alvarez, F.X.: From kinetic to collective behavior in thermal transport on semiconductors and semiconductor nanostructures. J. Appl. Phys. **115**, 164314 (2014)
16. De Tomas, C., Cantarero, A., Lopeandia, A.F., Alvarez, F.X.: Thermal conductivity of group-IV semiconductors from a kinetic-collective model. Proc R. Soc. A **470**, 20140371 (12 pp.) (2014)
17. Dreyer, W., Struchtrup, H.: Heat pulse experiments revisited. Contin. Mech. Thermodyn. **5**, 3–50 (1993)
18. Fujii, M., Zhang, X., Xie, H., Ago, H., Takahashi, K., Ikuta, T., Abe, H., Shimizu, T.: Measuring the thermal conductivity of a single carbon nanotube. Phys. Rev. Lett. **95**, 065502 (4 pp.) (2005)
19. Geim, A.K., Novoselov, K.S.: The rise of graphene. Nat. Mater. **6**, 183–191 (2007)
20. Ghosh, S., Calizo, I., Teweldebrhan, D., Pokatilov, E.P., Nika, D.L., Balandin, A.A., Bao, W., Miao, F., Lau, C.N.: Extremely high thermal conductivity of graphene: prospects for thermal management applications in nanoelectronic circuits. Appl. Phys. Lett. **92**, 151911 (3 pp.) (2008)
21. Gourley, S.A.: Travelling front solutions of a nonlocal Fisher equation. J. Math. Biol. **3**, 272–284 (2000)
22. Guo, Z.-Y., Hou, Q.-W.: Thermal wave based on the thermomass model. J. Heat Transf. Trans. ASME **132**, 072403 (6 pp.) (2010)
23. Hone, J., Whitney, M., Piskoti, C., Zettl, A.: Thermal conductivity of single-walled carbon nanotubes. Phys. Rev. B **59**, R2514–R2516 (1999)
24. Joseph, D.D.: Stability of Fluid Motion. Springer, New York (1976)
25. Joseph, D.D., Preziosi, L.: Heat waves. Rev. Mod. Phys. **61**, 41–73 (1989)
26. Jou, D., Casas-Vázquez, J., Lebon, G.: Extended irreversible thermodynamics revisited (1988–1998). Rep. Prog. Phys. **62**, 1035–1142 (1999)
27. Jou, D., Casas-Vázquez, J., Lebon, G., Grmela, M.: A phenomenological scaling approach for heat transport in nano-systems. Appl. Math. Lett. **18**, 963–967 (2005)
28. Jou, D., Casas-Vázquez, J., Lebon, G.: Extended Irreversible Thermodynamics, 4th revised edn. Springer, Berlin (2010)
29. Ju, Y.S., Goodson, K.E.: Phonon scattering in silicon films with thickness of order 100 nm. Appl. Phys. Lett. **74**, 3005–3007 (1999)
30. Lebon, G., Jou, D., Casas-Vázquez, J.: Understanding Nonequilibrium Thermodynamics. Springer, Berlin (2008)

31. Lee, S., Broido, D., Esfarjani, K., Chen, G.: Hydrodynamic phonon transport in suspended graphene. Nat. Commun. **6**, 6290 (9 pp.) (2015)
32. Levermore, C.D.: Relating Eddington factors to flux limiters. J. Quant. Spectrosc. Radiat. Transf. **31**, 149–160 (1984)
33. Levermore, C.D., Pomraning, G.C.: A flux-limited diffusion theory. Astrophys. J. **248**, 321–334 (1981)
34. Li, B., Wang, L., Casati, G.: Thermal diode: rectification of heat flux. Phys. Rev. Lett. **93**, 184301 (4 pp.) (2004)
35. Li, N., Ren, J., Wang, L., Zhang, G., Hänggi, P., Li, B.: Colloquium: phononics: manipulating heat flow with electronic analogs and beyond. Rev. Mod. Phys. **84**, 1045–1066 (2012)
36. Mendez, V., Fedotov, S., Horsthemke, W.: Reaction-Transport Systems. Mesoscopic Foundations, Fronts, and Spatial Instabilities. Springer, Berlin (2010)
37. Mihalas, D., Mihalas, B.W.: Foundations of Radiation Hydrodynamics. Oxford University Press, Oxford (1984)
38. Müller, I., Ruggeri, T.: Rational Extended Thermodynamics, 2nd edn. Springer, New York (1998)
39. Nika, D.L., Ghosh, S., Pokatilov, E.P., Balandin, A.A.: Lattice thermal conductivity of graphene flakes: comparison with bulk graphite. Appl. Phys. Lett. **94**, 203103 (3 pp.) (2009)
40. Nika, D.L., Pokatilov, E.P., Askerox, A.S., Balandin, A.A.: Phonon thermal conduction in graphene: role of Umklapp and edge roughness scattering. Phys. Rev. B **79**, 155413 (12 pp.) (2009)
41. Peierls, R.E.: Quantum Theory of Solids. Oxford University Press, London (1955)
42. Segal, D., Nitzan, A.: Spin-boson thermal rectifier. Phys. Rev. Lett. **94**, 034301 (4 pp.) (2005)
43. Sellitto, A., Alvarez, F.X., Jou, D.: Second law of thermodynamics and phonon-boundary conditions in nanowires. J. Appl. Phys. **107**, 064302 (7 pp.) (2010)
44. Sellitto, A., Alvarez, F.X., Jou, D.: Temperature dependence of boundary conditions in phonon hydrodynamics of smooth and rough nanowires. J. Appl. Phys. **107**, 114312 (7 pp.) (2010)
45. Sellitto, A., Cimmelli, V.A., Jou, D.: Analysis of three nonlinear effects in a continuum approach to heat transport in nanosystems. Physica D **241**, 1344–1350 (2012)
46. Sellitto, A., Jou, D., Bafaluy, J.: Nonlocal effects in radial heat transport in silicon thin layers and graphene sheets. Proc. R. Soc. A **468**, 1217–1229 (2012)
47. Sellitto, A., Cimmelli, V.A., Jou, D.: Entropy flux and anomalous axial heat transport at the nanoscale. Phys. Rev. B **87**, 054302 (7 pp.) (2013)
48. Shvarts, D., Delettrez, J., McCrory, L.R., Verdon, C.P.: Self-consistent reduction of the Spitzer-Härm electron thermal heat flux in steep temperature gradients in laser-produced plasmas. Phys. Rev. Lett. **47**, 247–250 (1981)
49. Terraneo, M., Peyrard, M., Casati, G.: Controlling the energy flow in nonlinear lattices: a model for a thermal rectifier. Phys. Rev. Lett. **88**, 094302 (4 pp.) (2002)
50. Tzou, D.Y.: Nonlocal behavior in phonon transport. Int. J. Heat Mass Transf. **54**, 475–481 (2011)
51. Tzou, D.Y., Guo, Z.-Y.: Nonlocal behavior in thermal lagging. Int. J. Therm. Sci. **49**, 1133–1137 (2010)
52. Wang, M., Guo, Z.-Y.: Understanding of temperature and size dependences of effective thermal conductivity of nanotubes. Phys. Lett. A **374**, 4312–4315 (2010)
53. Wang, M., Shan, X., Yang, N.: Understanding length dependences of effective thermal conductivity of nanowires. Phys. Lett. A **376**, 3514–3517 (2012)
54. Ward, A., Broido, D.A.: Intrinsic phonon relaxation times from first-principles studies of the thermal conductivities of Si and Ge. Phys. Rev. B **80**, 085205 (5 pp.) (2010)
55. Yang, N., Zhang, G., Li, B.: Carbon nanocone: a promising thermal rectifier. Appl. Phys. Lett. **93**, 243111 (3 pp.) (2008)

Chapter 6
Heat Transport with Phonons and Electrons and Efficiency of Thermoelectric Generators

Alternative power sources based on energy harvesting are promising candidates to substitute batteries due to their ability to extract power from the environment or secondary processes, as well as to attain fully autonomous systems without periodical human intervention. Thermoelectric energy harvesters have received special attention in recent years, due to the large amount of residual heat yielding from the current energy generation technology based on fossil fuels and from solar heating. Thermoelectric devices offer an attractive source of energy since they do not have moving parts, do not create pollution, do not make noise.

In practical applications, the definition of a good thermoelectric device is usually related to the dimensionless product ZT, with T being the operating temperature, and Z the so-called figure-of-merit, defined as

$$Z = \frac{\epsilon^2 \sigma_e}{\lambda} \tag{6.1}$$

wherein ϵ is the Seebeck coefficient, and σ_e the electrical conductivity. Although the thermal conductivity λ of thermoelectric materials is usually dominated by that of electrons, in several cases the lattice heat conductivity, due to phonons, has to be added [37]. For instance, in small systems (nanowires and nanoribbons with characteristic sizes of the order of 100 nm, or less) the electronic part of the thermal conductivity drops faster than the phononic contribution [61]. Other examples are some epitaxial superconducting films with high critical temperature, where the phonons contribution is 50–70 % of thermal conduction near the critical temperature [16, 21]. Therefore, in the denominator of Eq. (6.1) the total thermal conductivity is such that

$$\lambda = \lambda_p + \lambda_e \tag{6.2}$$

© Springer International Publishing Switzerland 2016
A. Sellitto et al., *Mesoscopic Theories of Heat Transport in Nanosystems*,
SEMA SIMAI Springer Series 6, DOI 10.1007/978-3-319-27206-1_6

with λ_p being the phonon contribution to the thermal conductivity, and λ_e the electron contribution to it [21, 58, 61].

Since the higher ZT, the higher the efficiency of a thermoelectric device,[1] in order to widen the applications of thermoelectric power generators, in the last decades there has been much research to improve the values of ZT beyond that of bulk materials, which show a low efficiency. However, ZT has remained approximately equal to 1 for the past several decades in the case of archetype materials at all temperature ranges. These materials include antimony telluride (Sb_2Te_3) and bismuth telluride (Be_2Te_3) for applications at room temperature, lead telluride (PbTe) at moderate temperatures, and silicon-germanium (SiGe) alloys at high temperatures.

One of the primary challenges in developing advanced thermoelectric materials is increasing the power factor $\epsilon^2 \sigma_e$ and reducing the thermal conductivity λ. To do so, it is needed decoupling ϵ, σ_e and λ, which are indeed typically strongly interdependent, in such a way that an increase in ϵ usually results in a decrease in σ_e, and a decrease in σ_e produces a decrease in the electronic contribution to λ, following from the Wiedemann-Franz law.

However, if the characteristic dimension of the material (or of the system) is reduced, the new variable *length scale* becomes available for the control of its properties. In particular, as the system size decreases and approaches to the nanometer scale, new opportunities are allowed to vary quasi-independently the aforementioned parameters. Nanomaterials, therefore, provide an interesting avenue to obtain more performing thermoelectric devices, for example, making nanocomposites, adding nanoparticles to a bulk material, or using one-dimensional nanostructures [7, 33]. Carbon nanotubes and graphene sheets as thermoelectric materials also exhibit improved thermoelectric properties. Nanosystems also offer the possibility of an additional control of the different transport coefficients [34, 40, 64, 65]. For instance, whenever the characteristic size of the system is comparable to the mean-free path of the different heat carriers (phonons, electrons, holes, etc.) it is known that a thermal-conductivity reduction can be realized over a wide temperature range, or the power factor (i.e., $\epsilon^2 \sigma_e$) can be increased at the same time by increasing ϵ more than σ_e is decreased [18, 43].

Although from experimental evidences it is clear the importance of using nanotechnologies in thermoelectricity, the search of a *very good* thermoelectric device is still far from its final solution. This principally because the physics at nanoscale in thermoelectric materials still presents several dark points, as for instance the exact role played by nonlocal and nonlinear effects [34, 40, 63, 64, 67]. Thus, the analysis of both effects may be useful to investigate new strategies for the optimization of thermoelectric devices, beside contributing to a deeper scientific understanding of thermoelectric effects.

---

[1]Refer to Eq. (6.31) in Sect. 6.1 for its explicit expression in the classical version in the case of a thermoelectric energy generator.

In the classical form, thermoelectric effects are described by the following constitutive equations

$$\mathbf{q} = -\lambda \nabla T + \left(\Pi + \frac{\mu_e}{z_e} \right) \mathbf{I} \tag{6.3a}$$

$$\mathbf{I} = -\sigma_e \epsilon \nabla T + \sigma_e \left[\mathbf{E} - \nabla \left(\frac{\mu_e}{z_e} \right) \right] \tag{6.3b}$$

wherein Π is the Peltier coefficient, $\mathbf{I}$ the electric-current density, $\mathbf{E}$ the electric field, μ_e the chemical potential of the electrons, and z_e the electric charge per unit mass of the electrons (which should be not confused with the electric charge of the carriers ϱ_e to which it is related as $\varrho_e = c_e z_e$, with c_e being mass fraction of electrons).

From Eq. (6.3a) it is possible to see that, even in absence of a temperature gradient, a heat flux may be generated due to an electrical current: this is the so-called the Peltier effect. The coefficient $\Pi = \pm (|q| / |I|)_{\Delta T=0}$ measures the amount of heat absorbed (or rejected) at the junction of two conductors of different materials kept at uniform temperature and crossed by an electric current of unit density [40]. Equation (6.3b) shows the property that, even in absence of an electric current, an electrical field $\mathbf{E}$ can be created by a temperature difference: this is known as Seebeck effect. The Seebeck coefficient ϵ (also known as thermoelectric power coefficient) measures the electrical potential produced by a unit temperature difference, in absence of electric current [40].

In the present chapter we study how to obtain enhanced versions of Eqs. (6.3), which have meaningful consequences on some well-known classical theoretical results, and in practical applications, i.e., on the efficiency η of thermoelectric energy generators. To this aim, we assume a cylindrical thermoelectric nanodevice whose longitudinal length L is much larger than the characteristic size of the transversal section. In this case we may represent it as a one-dimensional system and consider only one Cartesian component, namely, the longitudinal one y. In steady states we assume that the hot side of this system (i.e., that at $y = L$) is kept at the temperature T^h, and its cold side (i.e., that at $y = 0$) at the temperature T^c. Moreover, we suppose that both an electric current $\mathbf{I}$, and a quantity of heat per unit time enter uniformly into the hot side of the device and flow through it. In such a case the thermoelectric efficiency reads

$$\eta = \frac{P_{el}}{\dot{Q}_{tot}} \tag{6.4}$$

wherein P_{el} is the electric-power output, and $\dot{Q}_{tot}$ is the total heat supplied per unit time to the system.

All the results derived in the present chapter lie on the basic assumption that in thermoelectric materials the local heat flux $\mathbf{q}$ has two different contributions: the phonon partial heat flux $\mathbf{q}^{(p)}$ and the electron partial heat flux $\mathbf{q}^{(e)}$. In the simplest situation, those contributions are such that

$$\mathbf{q} = \mathbf{q}^{(p)} + \mathbf{q}^{(e)}. \qquad (6.5)$$

6.1 Two-Temperature Model for Thermoelectric Effects

In the present section we generalize Eqs. (6.3) whenever the different heat carriers (i.e., the phonons and the electrons in our case) no longer have the same temperature. Note that accounting for two different temperatures may be important, for instance, in the following physical situations

1. **Time-dependent situations: fast laser pulses**. When a laser pulse hits the surface of a system, initially the electrons capture the main amount of the incoming energy with respect to the phonons. Subsequently, through electron-phonon collisions, they give a part of it to the phonons. This may be of interest, for example, in the Raman thermometry (which is often utilized to measure the temperature in small electronic devices) or in information recording on optical discs (CD, DVD, Blu-Ray).
2. **Steady-state situations: nonequilibrium temperatures**. As the electron mean-free path ℓ_e is usually shorter than the phonon mean-free path ℓ_p, in heat propagation and when the longitudinal distance y is such that $\ell_e < y < \ell_p$, it is expected a very high number of electron collisions, and only scant phonon collisions. This yields that the electron temperature may reach its local-equilibrium value, whereas the phonon temperature may be still far from its own local-equilibrium value.
3. **Hot electrons**. When the electron mean-free path corresponding to electron-phonon collisions is long, one may have the so-called "hot electrons", namely, a population of electrons whose average kinetic energy (i.e., the kinetic temperature) is considerably higher than that of the phonons [2].

Since we assume that the heat carriers behave as a mixture of gases flowing through the crystal lattice [11], it seems logical to suppose that the internal energy of phonons per unit volume u_p, the internal energy of electrons per unit volume u_e and the electrical charge per unit volume of electrons ϱ_e belong to the state space. In particular, we assume that those state-space variables are ruled by the following evolution equations:

$$\dot{u}_p = -\nabla \cdot \mathbf{q}^{(p)} \qquad (6.6a)$$

$$\dot{u}_e = -\nabla \cdot \mathbf{q}^{(e)} + \mathbf{E} \cdot \mathbf{I} \qquad (6.6b)$$

$$\dot{\varrho}_e = -\nabla \cdot \mathbf{I}. \qquad (6.6c)$$

If the total internal energy per unit volume of the system u is supposed to be given by the constitutive relation

$$u = u_p + u_e \tag{6.7}$$

and Eq. (6.5) is taken into account, then the summation of Eqs. (6.6a) and (6.6b) turns out the well-known energy-balance equation

$$\dot{u} + \nabla \cdot \mathbf{q} = \mathbf{E} \cdot \mathbf{I}$$

obtained in Ref. [15] in the absence of a magnetic field. According with the basic principles of EIT [34, 40], we may assume that the fluxes of the previous unknown variables (namely, $\mathbf{q}^{(p)}$, $\mathbf{q}^{(e)}$ and $\mathbf{I}$) are the other state-space variables. Whenever the relaxation times of those fluxes are negligible, the entropy per unit volume s is such that $s = s\left(u_p, u_e, \varrho_e\right)$ [59], and from the Gibbs relation we have

$$ds = \left(\frac{\partial s}{\partial u_p}\right) du_p + \left(\frac{\partial s}{\partial u_e}\right) du_e + \left(\frac{\partial s}{\partial \varrho_e}\right) d\varrho_e \Rightarrow$$

$$\dot{s} = \left(\frac{1}{T_p}\right) \dot{u}_p + \left(\frac{1}{T_e}\right) \dot{u}_e - \left(\frac{\mu_e}{z_e T_e}\right) \dot{\varrho}_e \tag{6.8}$$

wherein $T_p = \left(\partial s/\partial u_p\right)^{-1}$ is the phonon temperature, $T_e = \left(\partial s/\partial u_e\right)^{-1}$ is the electron temperature, and $\mu_e/\left(z_e T_e\right) = -\partial s/\partial \varrho_e$. From the theoretical point of view, we may postulate the following further constitutive equations which relate the partial internal energies appearing in Eqs. (6.6) to those temperatures:

$$u_p = c_v^{(p)} T_p \tag{6.9a}$$

$$u_e = c_v^{(e)} T_e \tag{6.9b}$$

wherein $c_v^{(p)}$ and $c_v^{(e)}$ are the phonon and the electron specific heats at constant volume [41], respectively. Since the total internal energy u can be expressed through the average temperature T as $u = c_v T$, being $c_v = c_v^{(p)} + c_v^{(e)}$ the specific heat at constant volume of the whole system [3], from the coupling of Eqs. (6.7) and (6.9) we obtain

$$T = \frac{c_v^{(p)} T_p + c_v^{(e)} T_e}{c_v} \tag{6.10}$$

which states a link between T_p, T_e and T, the latter being a measurable quantity in practical applications. Note that in the very general case $c_v^{(p)}$ and $c_v^{(e)}$ should be temperature-dependent functions, but here we deal only with the simplest situation in which those material functions are constant, in order to emphasize the essential physical ideas and their consequences. In other words, since we regard the phonons and electrons as a mixture of gases flowing through the crystal lattice [1, 11], each of which is endowed with its own temperature, according with the theory of fluid mixtures with different temperatures [22, 49–51], we assume that each constituent obeys the same balance laws as a single fluid. The average temperature of the mixture has been introduced by the consideration that the internal energy of the mixture is the same as in the case of a single-temperature mixture [50, 51].

The substitution of Eqs. (6.6) into Eq. (6.8) leads to

$$\dot{s} = -\frac{\nabla \cdot \mathbf{q}^{(p)}}{T_p} - \frac{\nabla \cdot \mathbf{q}^{(e)}}{T_e} + \left(\frac{\mu_e}{z_e T_e}\right) \nabla \cdot \mathbf{I} + \frac{\mathbf{E} \cdot \mathbf{I}}{T_e} =$$

$$-\nabla \cdot \left(\frac{\mathbf{q}^{(p)}}{T_p} + \frac{\mathbf{q}^{(e)}}{T_e} - \frac{\mu_e}{z_e T_e}\mathbf{I}\right)$$

$$-\frac{\mathbf{q}^{(p)} \cdot \nabla T_p}{T_p^2} - \frac{\mathbf{q}^{(e)} \cdot \nabla T_e}{T_e^2} - \frac{\mathbf{I}}{T_e}\nabla\left(\frac{\mu_e}{z_e}\right) + \left(\frac{\mu_e}{z_e T_e^2}\right)\mathbf{I} \cdot \nabla T_e + \frac{\mathbf{E} \cdot \mathbf{I}}{T_e}.$$

$$(6.11)$$

Recalling that the time rate of the entropy density has to obey the balance law (1.9), its comparison with Eq. (6.11) leads to the following identifications:

$$\mathbf{J}^s = \frac{\mathbf{q}^{(p)}}{T_p} + \frac{\mathbf{q}^{(e)}}{T_e} - \frac{\mu_e}{z_e T_e}\mathbf{I} \qquad (6.12a)$$

$$\sigma^s = -\frac{1}{T_p}\left\{\mathbf{q}^{(p)} \cdot \frac{\nabla T_p}{T_p}\right\} - \frac{1}{T_e}\left\{\left[\mathbf{q}^{(e)} - \frac{\mu_e}{z_e}\mathbf{I}\right] \cdot \frac{\nabla T_e}{T_e}\right\} + \frac{1}{T_e}\left\{\mathbf{I} \cdot \left[\mathbf{E} - \nabla\left(\frac{\mu_e}{z_e}\right)\right]\right\}$$

$$= \sum_\alpha \mathbf{J}^\alpha \cdot \mathbf{X}^\alpha \qquad (6.12b)$$

wherein $\mathbf{J}^\alpha$ is the thermodynamic flux, and $\mathbf{X}^\alpha$ is its conjugated thermodynamic force [17, 40]. As we observed in Chap. 1, experience indicates that for a large class of irreversible processes, the thermodynamic fluxes can be regarded as linear functions of the forces, to a good approximation [13, 17, 23, 34]. This choice, which is also the simplest way to ensure that σ^s is a non-negative quantity whatever the thermodynamic process is, allows us to write the following phenomenological relations for the fluxes appearing in our model:

$$-\mathbf{q}^{(p)} = L_{11}\frac{\nabla T_p}{T_p} + L_{12}\frac{\nabla T_e}{T_e} + L_{13}\left[\mathbf{E} - \nabla\left(\frac{\mu_e}{z_e}\right)\right] \qquad (6.13a)$$

$$\left(\frac{\mu_e}{z_e}\right)\mathbf{I} - \mathbf{q}^{(e)} = L_{21}\frac{\nabla T_p}{T_p} + L_{22}\frac{\nabla T_e}{T_e} + L_{23}\left[\mathbf{E} - \nabla\left(\frac{\mu_e}{z_e}\right)\right] \qquad (6.13b)$$

$$\mathbf{I} = L_{31}\frac{\nabla T_p}{T_p} + L_{32}\frac{\nabla T_e}{T_e} + L_{33}\left[\mathbf{E} - \nabla\left(\frac{\mu_e}{z_e}\right)\right]. \qquad (6.13c)$$

As the thermoelectric effects arise from the physical interrelation between heat flow and electric current, in the classical thermoelectric models the cross-coefficients $L_{\alpha\beta}$ (with $\alpha \neq \beta$) are only related to the coupled transport of heat and electricity. In the present model, instead, the thermoelectric effect is driven by three generalized thermodynamic forces, namely, the phonon and electron temperature gradients, and the force due to the electric field and to the chemical potential of the electric charge. As a consequence, the cross-coefficients L_{12} and L_{21} account for the cross effects due to the different temperatures of the heat carriers (which are lacking in the classical case), the coefficients L_{13} and L_{31} represent the coupling between the phonon temperature gradient and the electric current, and the coefficients L_{23} and L_{32} account for the coupling between the electron temperature gradient and the electric current. In the standard thermoelectric models, with $T_p = T_e$, we have only two cross-coefficients and a 2×2 transport matrix. Here, instead, we have six cross-coefficients and a 3×3 transport matrix. To be compatible with the second law of thermodynamics (expressed as the positive-definite character of the entropy production) the matrix of the phenomenological coefficients $L_{\alpha\beta}$ must be positive definite [13, 17]. On the other hand, the Onsager reciprocal relations (OR) [44, 45] ensure that the transport matrix is symmetric, so that the Sylvester criterion concerning the positive definiteness of real-symmetric $n \times n$ matrices is applicable. Such a criterion states that the positiveness of all the leading principal minors of the matrix is necessary and sufficient to ensure that it is positive definite [20].

By means of Eq. (6.5), in a somewhat different form the phenomenological equations (6.13) may be also written as

$$\mathbf{q} = -\left(\lambda_p + \lambda_{ep}\right)\nabla T_p - \left(\lambda_e + \lambda_{pe}\right)\nabla T_e + \left(\frac{\mu_e}{z_e} + \Pi\right)\mathbf{I} \qquad (6.14a)$$

$$\mathbf{I} = -\sigma_e\epsilon\nabla T_e + \sigma_e\left[\mathbf{E} - \nabla\left(\frac{\mu_e}{z_e}\right)\right] \qquad (6.14b)$$

once the following identifications are made:

$$\begin{cases} L_{11} = \lambda_p T_p & L_{12} = \lambda_{pe} T_e & L_{13} = 0 \\ L_{21} = \lambda_{ep} T_p & L_{22} = \lambda_e T_e + \sigma_e\varepsilon T_e\Pi & L_{23} = -\sigma_e\Pi \\ L_{31} = 0 & L_{32} = -\sigma_e\varepsilon T_e & L_{33} = \sigma_e \end{cases} \qquad (6.15)$$

with λ_{pe} and λ_{ep} expressing the contributions to the total thermal conductivity due to the possible interactions between the different heat carriers [59]. The use of these thermal conductivities, arising from the phenomenological coefficients L_{12} and L_{21} in Eqs. (6.13a) and (6.13b), respectively, are representative of the cross effects in the constitutive equations for the diffusive fluxes $\mathbf{q}^{(p)}$ and $\mathbf{q}^{(e)}$. However, as it will be seen in Sect. 6.1.2, these cross effects do not play any relevant role on the efficiency of thermoelectric energy conversion, being the difference in the two temperatures the principal responsible for possible enhancements of it. Moreover, in Eq. (6.15) we assumed $L_{13} = L_{31} = 0$ since the phonons are not expected to be directly sensitive to the external electric field, at least in a first approximation. For polar lattices, this possibility would be open, and L_{13} could be different from zero. However, here we take the simplest expression.

Owing to the OR [44, 45], from Eq. (6.15) we obtain

$$\lambda_{pe} = \lambda_{ep} \left(\frac{T_p}{T_e} \right) \tag{6.16a}$$

$$\Pi = \epsilon T_e. \tag{6.16b}$$

The relation (6.16b) is well known in thermoelectricity when T_e is replaced by T. In that case it is referred to as the *second Kelvin relation*. The different result predicted by Eq. (6.16b) with respect to the usual statement of the second Kelvin relation (i.e., $\Pi = \epsilon T$) should not be considered as a surprising and unexpected result. In fact, the classical statement is correct in the case that phonons and electrons have the same temperature. In the case they have different temperatures, instead, the relation (6.16b) seems more appropriate. That result is also not against OR [44, 45] but, on the contrary, it is an illustration of them, more accurate and precise than the classical expression with a single temperature, as it is obtained from the symmetry of coefficients $L_{\alpha\beta}$ appearing in Eq. (6.15).

Finally, according to the Sylvester's criterion, the additional constraints

$$\lambda_p > 0 \tag{6.17a}$$

$$\lambda_p \left(\lambda_e + \sigma_e \Pi \right) > \lambda_{pe} \lambda_{ep} \tag{6.17b}$$

$$\lambda_p \lambda_e > \lambda_{pe} \lambda_{ep} \tag{6.17c}$$

are necessary and sufficient to ensure that the entropy production is always non-negative. In fact, due the positive definiteness of the matrix of the transport coefficients, the entropy production is positive whenever at least one thermodynamic force is different from zero, and vanishes when all the thermodynamic forces are zero. Such a situation characterizes the quasi-static processes [66], which correspond to zero entropy production.

6.1.1 Possible Estimations of the Phonon and Electron Temperature

Since from the theoretical point of view it seems of interest to accounting for T_e and T_p, it would be useful in practical applications to find a way to estimate those temperatures [5, 55]. Equation (6.16b) may be important for this aim. In fact, by means of a usual thermometer, one is only able to check the average temperature T, defined by Eq. (6.10). Equation (6.16b) turns out that T_e should be given as the ratio between Π and ϵ. Therefore, if one is able to measure Π and ϵ then, by the coupling of Eqs. (6.10) and (6.16b), in principle, it is possible to determine T_p, too. In fact, setting

$$
\begin{cases}
\alpha = \dfrac{c_v^{(e)}}{c_v}, \; 1 - \alpha = \dfrac{c_v^{(p)}}{c_v} \\
\beta_1 = \dfrac{T_e}{T}, \; \beta_2 = \dfrac{T_p}{T}
\end{cases}
\tag{6.18}
$$

whereas from Eq. (6.10) we have

$$
\beta_2 = \frac{1}{1-\alpha} - \left(\frac{\alpha}{1-\alpha}\right)\beta_1.
\tag{6.19}
$$

It is worth to note that at nanoscale the different material functions may deviate from their corresponding bulk values, in such a way that several methods of measuring them may be found in literature [31, 62].

From the practical point of view, the way of having a truly different temperature for electrons (and holes) and lattice (phonons) is by means of a pulse-laser excitation yielding its energy to the charged particles (electrons), which later share their energy excess with the lattice. Despite the lattice may be also charged, the much higher mass of the ions makes that the electrons may absorb more energy from the pulse. However, this strategy is not of interest for practical thermoelectric devices, because it would require spending energy on the laser, from which only a part would be taken by the system. A different strategy would be by using a system composed of two (or several) thin layers (as for instance a superlattice) the characteristic dimensions of which are shorter than the phonon mean-free path, but larger than the electron mean-free path. If the electron and phonon contributions to the heat flux within the layers are, respectively, $\mathbf{q}^{(e)} = -\lambda_e \nabla T_e$ and $\mathbf{q}^{(p)} = -\lambda_p \nabla T_p$, and the temperature discontinuities at an interface are $\Delta T_e = R_e q_e$ and $\Delta T_p = R_p q_p$, R_e and R_p being the respective thermal resistance of the interface, one may obtain the profile for T_e and T_p along the system. Both temperatures must be equal to the heat baths at the two ends of the whole system, but they may be different from each other along the system. By using suitable materials to have R_e, R_p, λ_e and λ_p (or a suitably reduced effective phonon thermal conductivity), one could have regions with T_e higher than T_p, and regions with T_e lower than T_p. A detailed analysis of both profiles should be carried out to study the global effects of these differences on the efficiency of the thermoelectric conversion in the whole device. Of course, electrons exchange energy with phonons everywhere. Thus, it would be convenient that the electron heat flux is relatively large, in order that this energy exchange does not bring to zero the temperature differences. These effects should be studied in detail.

Indeed, in the present two-temperature model, the problem of measuring T_e and T_p may be also related to the phonon-drag phenomenon [8, 46]. The classical theory of thermoelectricity, in fact, is based on the assumption that the flow of charge carriers and phonons can be treated independently. Under this assumption, the Seebeck coefficient solely depends by the spontaneous electron diffusion. However, when the two flows are linked, the effect of electron-phonon scattering should be taken into account. Hence, in general, the Seebeck coefficient shows two independent contributions: the conventional electron-diffusion contribution and the phonon-drag contribution [46]. The diffusion part is caused by the spatial variation of the electronic occupation in the presence of a thermal gradient, whereas the drag part arises by the interaction between anisotropic lattice vibrations and mobile charge carriers. The overall phonon-drag effect leads to an increase in the Seebeck coefficient. If we look at Eq. (6.14b), we may observe that it allows to introduce the following effective Seebeck coefficient

$$\beta_1 = \frac{\epsilon_{\text{eff}}}{\epsilon} \qquad (6.20)$$

which also allows to claim that the deviation of the effective Seebeck coefficient from its bulk value represents a further possible measurement of the electron temperature. Moreover, along with previous observations about the phonon-drag phenomenon, from Eq. (6.20) we further claim that in general β_1 should be higher than unit, namely, $T_e > T_p$.

Finally, observing that $\alpha \in]0; 1[$, the physical constraint $\beta_2 > 0$ implies that $\beta_1 \in]0; \alpha^{-1}[$. From Eq. (6.19) it is easy to see that the condition $\beta_1 = 1 \Rightarrow \beta_2 = 1$, namely, in this case $T_e \equiv T_p \equiv T$: the two-temperature model described by Eqs. (6.14) reduces to the usual single-temperature model [40, 56], i.e., to Eqs. (6.3), once the thermal conductivity λ of the material is supposed to be given as $\lambda = \lambda_p + \lambda_e + 2\lambda_{pe}$.

6.1.2   Efficiency of a Thermoelectric Energy Generator in a Two-Temperature Model

Let us now investigate the consequences of accounting for two different temperatures for phonon and electrons on the efficiency of a nano-thermoelectric energy generator, as that previously introduced. For the sake of simplicity, let us suppose that $\mathbf{q}^{(p)}$, $\mathbf{q}^{(e)}$ and $\mathbf{I}$ are parallel to the nanowire, and that $\mathbf{E}$ takes a constant value on each of the planes orthogonal to the nanowire.

In these conditions, for vanishing values of μ_e/z_e, Eqs. (6.14) become

$$q = q^{(p)} + q^{(e)} = -\Lambda_p \nabla_y T_p - \Lambda_e \nabla_y T_e + \Pi I \qquad (6.21a)$$

$$I = -\sigma_e \epsilon \nabla_y T_e + \sigma_e E \qquad (6.21b)$$

wherein $\Lambda_p = \lambda_p + \lambda_{ep}$, and $\Lambda_e = \lambda_e + \lambda_{pe}$.

To proceed further, we explicitly observe that $\dot{Q}_{tot}$ may be obtained by integrating the heat flux on the section of the entry side of the nanowire. However, for very small sections, if A is the area of the cross section, we have

$$\lim_{A \to 0} \dot{Q}_{tot} = |q| .\tag{6.22}$$

The situation described above reflects just what happens in the nanowire considered here, so that we can substitute $|q|$ to $\dot{Q}_{tot}$ in Eq. (6.4). On the other hand, since we do not have any "a priori" information on the dependence of T_p and T_e on y, we rewrite Eq. (6.21a) under the approximation

$$\nabla_y T_p = \frac{T_p(L) - T_p(0)}{L} = \frac{T_p^h - T_p^c}{L}\tag{6.23a}$$

$$\nabla_y T_e = \frac{T_e(L) - T_e(0)}{L} = \frac{T_e^h - T_e^c}{L}\tag{6.23b}$$

so that it reads

$$\dot{Q}_{tot} \equiv |q| = \Lambda_p \frac{T_p^h - T_p^c}{L} + \Lambda_e \frac{T_e^h - T_e^c}{L} + \Pi I.\tag{6.24}$$

If, for the sake of simplicity, ϵ and σ_e are supposed to be constant, the electric-power output is equal to the work made by the Coulomb force per unit time for moving the electric charges from $y = L$ to $y = 0$, namely,

$$P_{el} = I \int_L^0 E\,dy = -I \int_0^L E\,dy = -I \int_0^L \left(\frac{I}{\sigma_e} + \epsilon \nabla_y T_e \right) dy$$

$$= -\frac{I^2}{\sigma_e} L - \epsilon [T_e(L) - T_e(0)] = -\frac{I^2}{\sigma_e} L + I\epsilon \left(T_e^h - T_e^c \right).\tag{6.25}$$

Inserting Eqs. (6.24) and (6.25) into Eq. (6.4) we obtain

$$\eta = \frac{I\epsilon \left(T_e^h - T_e^c \right) - I^2 L \sigma_e^{-1}}{\Lambda_p \left(T_p^h - T_p^c \right) L^{-1} + \Lambda_e \left(T_e^h - T_e^c \right) L^{-1} + \Pi I}.\tag{6.26}$$

If we introduce in Eq. (6.26) the coefficients β_1 and β_2 defined in Eqs. (6.18), and take into account Eq. (6.16b), by straightforward calculations we have

$$\eta = \left(1 - \frac{T^c}{T^h} \right) \left(\frac{\epsilon x - \dfrac{\lambda x^2}{\sigma_e \beta_1}}{\dfrac{\gamma + 1}{T^h} + \epsilon x} \right) = \eta_c \eta_r\tag{6.27}$$

wherein $\lambda = \Lambda_p + \Lambda_e$, $\gamma = (\Lambda_p/\lambda)(\beta_2/\beta_1 - 1)$, and x is the following ratio between the electric current and the heat flux:

$$x = \frac{IL}{\lambda\,(T^h - T^c)}. \tag{6.28}$$

In Eq. (6.27) $\eta_c = 1 - T^c/T^h$ is the usual Carnot efficiency, and η_r is a *reduced efficiency*. Although in Refs. [29, 30] it is shown that, in principle, even the Carnot efficiency is attainable, the actual devices have performances which are notoriously much more modest, that is, $\eta_r \ll 1$. Therefore, in practical applications one should find the right way to enhance η_r in order to have a good thermoelectric efficiency. Indeed, it is easy to see that whenever the ratio x defined above gets the value

$$x_{\text{opt}} = \left(\frac{\gamma + 1}{\epsilon T}\right)\left(\sqrt{1 + \frac{Z_{\text{eff}}T\beta_1}{\gamma + 1}} - 1\right) \tag{6.29}$$

with $Z_{\text{eff}} = \epsilon^2\sigma_e/\lambda$ as an effective figure-of-merit, then the reduced efficiency gets its maximum value, and the thermoelectric efficiency reads

$$\eta_{\max} = \eta_c\left[\frac{Z_{\text{eff}}T\beta_1 + 2(\gamma + 1)\left(1 - \sqrt{1 + \dfrac{Z_{\text{eff}}T\beta_1}{\gamma + 1}}\right)}{Z_{\text{eff}}T\beta_1}\right] \tag{6.30}$$

which reduces to the classical form for the maximum thermoelectric efficiency [4, 40]

$$\eta_{\max,\text{classic}} = \eta_c\left[\frac{ZT + 2\left(1 - \sqrt{1 + ZT}\right)}{ZT}\right] \tag{6.31}$$

whenever T_p and T_e coincide, i.e., when $\beta_1 = \beta_2 = 1$ and $\gamma = 0$. From Eq. (6.30) the usual result that the larger the figure-of-merit, the higher the efficiency of a thermoelectric device is also recovered.

However, Eq. (6.30) clearly points out that also the differences between T_p and T_e influence $\eta_{\max}$. The cross effects related to the phenomenological coefficients L_{12} and L_{21} in Eqs. (6.13), instead, do not play any role in the efficiency. This may be interesting in practical applications, since most of the research is focusing only on the search of new materials with high values of the figure-of-merit.

In Fig. 6.1 we plot the behavior of the ratio $\eta_{\max}/\eta_c$ as a function of β_1 for two different values of the nondimensional parameter α, i.e., $\alpha = 0.05$ and $\alpha = 0.75$.

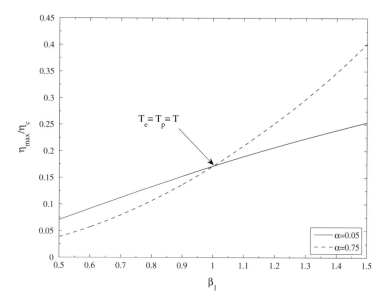

Fig. 6.1 Behavior of η_{max}/η_c versus β_1 for two different values of the nondimensional parameter $\alpha = c_v^{(e)}/c_v$: theoretical results arising from Eq. (6.30). $\beta_1 < 1 \Rightarrow T_p > T_e$, $\beta_1 = 1 \Rightarrow T_p = T_e$, and $\beta_1 > 1 \Rightarrow T_p < T_e$. For the sake of computation, we assumed $Z_{eff} = 1$

For the sake of illustration we assumed $\Lambda_p = \Lambda_e$, as the materials commonly used in thermoelectric applications show a phonon thermal conductivity which is approximately equal to the electron thermal conductivity. Moreover, we supposed that $Z_{eff}T = 1$.

As it can be seen, the maximum efficiency increases for increasing values of β_1. This means that the higher T_e with respect to T_p, the better the performances of thermoelectric devices. Indeed, Fig. 6.1 also allows to analyze the role played by $c_v^{(e)}$ and $c_v^{(p)}$, the latter being usually higher than the former. In fact, it points out that the (positive) slope of the curve $\eta_{max} = \eta_{max}(\beta_1)$ gradually decreases whenever α assumes small values, whereas it basically takes a constant value when α reaches high enough values, in such a way that whenever $T_e > T_p$, the higher α, the higher η_{max}.

At the very end, we observe that in Fig. 6.1 the value $\eta_{max}/\eta_c = 0.17$, attained whenever $\beta_1 = 1$ corresponds to the case of a single-temperature model.[2] Thus, whenever the electron temperature is higher than the phonon temperature, the two-temperature model yields an efficiency which is higher than that of the usual single-temperature model.

---

[2] Along with previous observations in this case one also has $\beta_2 = 1$, and moreover $\gamma = 0$.

6.1.3 Influence of the Electric-Charge Density on the Optimal Efficiency of a Thermoelectric Energy Generator

In recent years "functionally graded materials" (FGMs) [6], i.e., a new class of advanced materials with varying properties over a changing dimension, are attracting the attention of many research groups. In FGMs the properties change continuously, or quasi continuously, along one direction, and this implies that the different material functions may be assumed to be continuous, or quasi-continuous. Their versatility allows the use of these materials in thermoelectric applications, too. In particular, the efficiency of thermoelectric devices can be improved by adjusting the carriers' concentration along the material's length. This can be achieved by employing a functionally graded thermoelectric material (FGTM), with the carriers' concentration optimized for operating over a specific temperature range [32, 38, 39, 42].

Therefore, it is important to go deeper in the analysis of the thermoelectric efficiency, and account not only for the effects due to the different temperatures, but also for those due to the term $\mu_e c_e / \varrho_e$, which we previously neglected.

Recalling that when the relaxation times of the diffusive fluxes are vanishingly small, the entropy per unit volume s only depends on the unknown variables u_p, u_e and ϱ_e [59], from the definition of chemical potential, it follows

$$\frac{\mu_e c_e}{\varrho_e} = f\left(u_p, u_e, \varrho_e\right) \tag{6.32}$$

which allows to rewrite Eq. (6.14b) as

$$\mathbf{I} = -\sigma_e \widetilde{\epsilon} \nabla T_e - \left(\sigma_e \epsilon \frac{\partial f}{\partial T_p}\right) \nabla T_p + \sigma_e \left(\mathbf{E} - \frac{\partial f}{\partial \varrho_e} \nabla \varrho_e\right) \tag{6.33}$$

wherein $\widetilde{\epsilon} = \epsilon + \partial f / \partial T_e$.

For the sake of simplicity, in the next we approximate the material parameters $\widetilde{\epsilon}$, σ_e, $\partial f / \partial T_p$ and $\partial f / \partial \varrho_e$ by their mean values $\overline{\epsilon}$, $\overline{\sigma_e}$, $\overline{f_p}$ and $\overline{f_\varrho}$ in $[0, L]$. This mathematical assumption, which allows our calculations to be reduced to a simpler level, is tantamount to suppose that those constitutive quantities suffer only very small variations in $[0, L]$. Moreover, owing to the reduced length, we can approximate all the gradients in Eq. (6.33) by their measured values $\Delta T_p / L$, $\Delta T_e / L$ and $\Delta \varrho_e / L$, with $\Delta T_e = T_e^h - T_e^c$, $\Delta T_p = T_p^h - T_p^c$, and $\Delta \varrho_e = \varrho_e^h - \varrho_e^c$. In this case, Eqs. (6.24) and (6.25) change in

$$|q| = \Lambda_p \frac{\Delta T_p}{L} + \Lambda_e \frac{\Delta T_e}{L} + \left(\Pi + \overline{f}\right) I \tag{6.34a}$$

$$P_{\text{el}} = -\frac{I^2 L}{\overline{\sigma_e}} + \overline{\epsilon} I \Delta T_e + I \overline{f_p} \Delta T_p - I \left|\overline{f_\varrho}\right| \Delta \varrho_e. \tag{6.34b}$$

By observing that in a one-dimensional system, when a conductive material is subjected to a thermal gradient, the charge carriers migrate from the hotter side to the colder one, and, in the open-circuit condition, they accumulate in the cold region, generating so an electric potential difference, we conclude that the electric-current density $\mathbf{I}$ and the temperature gradients ∇T_e and ∇T_p have to be opposite vectors for arbitrary values of the material coefficients $\sigma_e \widetilde{\epsilon}$ and $(\sigma_e \epsilon) \, \partial f / \partial T_p$. As a consequence, by Eq. (6.33) we infer that $\partial f / \partial T_e > 0$ and $\partial f / \partial T_p > 0$. We note that in principle the Seebeck coefficient may assume either a positive value (for example in the case of a p-type semiconductor wherein the current is brought by holes), or a negative one (for example in the case of a n-type semiconductor wherein the current is carried by electrons). For the sake of illustration, here we analyze the case of a positive Seebeck coefficient. Moreover, since the electric current is always parallel and concordant to the gradient of the electric charge, still by Eq. (6.33) we conclude that $\partial f / \partial \varrho_e < 0$. By substituting Eqs. (6.34) into Eq. (6.4) we arrive to the following expression of the thermoelectric efficiency

$$\eta = \frac{-\dfrac{I^2 L}{\overline{\sigma_e}} + \overline{\epsilon} I \Delta T_e + I \overline{f_p} \Delta T_p - I \left| \overline{f_\varrho} \right| \Delta \varrho_e}{\Lambda_p \dfrac{\Delta T_p}{L} + \Lambda_e \dfrac{\Delta T_e}{L} + \left(\Pi + \overline{f} \right) I}. \tag{6.35}$$

The use of the variable x defined in Eq. (6.28) and of the new variable

$$w = \sqrt{\frac{\Delta \varrho_e}{IL}} \tag{6.36}$$

allows to rewrite Eq. (6.35) as [48]

$$\eta = \eta_c \left[\frac{-\left(\dfrac{\lambda}{\overline{\sigma_e} \beta_1} \right) x^2 + \dfrac{\Gamma}{\beta_1} x - \left(\dfrac{\lambda^2}{\beta_1} \left| \overline{f_\varrho} \right| \right) x^2 w^2}{\dfrac{\gamma + 1}{T^h} + \overline{\epsilon} \left(1 + \dfrac{\delta}{\beta_1} \right) x} \right] \tag{6.37}$$

wherein $\Gamma = \overline{\epsilon} \beta_1 + \beta_2 \overline{f_p}$, $\gamma = \Lambda_p \lambda^{-1} \left(\beta_2 \beta_1^{-1} - 1 \right)$, and $\delta = \overline{f} / \left(\overline{\epsilon} T^h \right)$. If we introduce an effective figure-of-merit defined as

$$Z_{\text{eff}} = \frac{(\overline{\epsilon})^2 \overline{\sigma_e}}{\lambda}$$

then, in correspondence of the following values of the variables above,

$$x_{\text{opt}} = \frac{\gamma + 1}{\overline{\epsilon} T \left(1 + \delta \beta_1^{-1} \right)} \left[-1 + \sqrt{1 + \left(\dfrac{\Gamma}{\overline{\epsilon}} \right) \left(\dfrac{Z_{\text{eff}} T}{\gamma + 1} \right) \left(1 + \dfrac{\delta}{\beta_1} \right)} \right] \tag{6.38a}$$

$$w_{\text{opt}} = 0 \tag{6.38b}$$

Eq. (6.37) gets its maximum [48], which reads

$$\eta_{max} = \frac{\eta_c (\gamma + 1)}{\beta_1 \left(1 + \delta\beta_1^{-1}\right)^2} \left[\frac{2}{Z_{eff}T} + \frac{\Gamma \left(1 + \delta\beta_1^{-1}\right)}{\bar{\epsilon} (\gamma + 1)} - \frac{2}{Z_{eff}T} \sqrt{1 + \frac{\Gamma \left(1 + \delta\beta_1^{-1}\right)}{\bar{\epsilon} (\gamma + 1)} Z_{eff}T} \right].$$

(6.39)

Equation (6.39), which generalizes the results of previous section under the hypothesis that the terms proportional to $\mu_e c_e / \varrho_e$ are negligible in Eqs. (6.14), points out that the optimum efficiency strongly depends not only on the two temperatures T_p and T_e, but also on the gradient of the charge carriers.

In particular, from Eq. (6.39) it is easy to see that

$$Z_{eff}T \rightarrow \infty \Rightarrow \eta \rightarrow \eta_c \left(\frac{\Gamma}{\bar{\epsilon}\beta_1}\right) \left(1 + \frac{\delta}{\beta_1}\right)^{-1}$$

so that the material functions $\tilde{\epsilon}$, $\partial f / \partial T_p$ and $\partial f / \partial \varrho_e$ have to fulfill the following physical restriction

$$\left(\frac{\Gamma}{\bar{\epsilon}\beta_1}\right) \left(1 + \frac{\delta}{\beta_1}\right)^{-1} \leq 1.$$

This consideration, joined with the further observation that, whenever the terms in f can be neglected in Eqs. (6.14), then $\eta \rightarrow \eta_c$ as $Z_{eff}T \rightarrow \infty$, allows to conclude that the presence of a non-vanishing gradient of the electric charges worsen the thermoelectric efficiency. This is logical, since the inhomogeneity in the charge density induces a current circulation inside the conductor and, as a consequence, an enhancement of dissipation by Joule effect.

For the sake of illustration, we evaluate f in correspondence of fixed values of T_e and T_p, so that $f = f(\varrho_e)$, and plot the ratio η_{max}/η_c in such a situation. Although in a special case, this simplifying assumption allows us to show in the next figures both the role played by the two temperatures, and by the gradient of electric charges.

In Fig. 6.2 we plot the behavior of the ratio η_{opt}/η_c as a function of β_1, for two different values of the nondimensional parameter α, i.e., $\alpha = 0.05$ and $\alpha = 0.75$, and when $\delta = 1$. For the sake of illustration, in our computation we assumed that $\Lambda_p = \Lambda_e$, as the materials commonly used in thermoelectric applications show a phonon thermal conductivity which is approximately equal to the electron thermal conductivity, and supposed $Z_{eff}T = 1$.

As it can be seen, in this case the maximum efficiency increases for increasing values of β_1. This means that the bigger T_e with respect to T_p, the better the performance of the thermoelectric device. Indeed, Fig. 6.2 also allows to analyze the role played by $c_v^{(e)}$ and $c_v^{(p)}$, the latter being usually higher than the former. In fact, it points out that the (positive) slope of the curve $\eta_{opt} = \eta_{opt}(\beta_1)$ gradually decreases whenever α assumes very small values, whereas it takes an almost constant value when α reaches high enough values, in such a way that whenever $T_e > T_p$, the

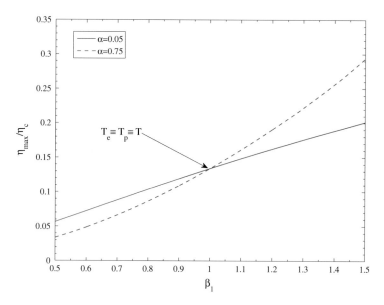

Fig. 6.2 Behavior of $\eta_{\max}/\eta_c$ versus β_1 for two different values of the non-dimensional parameter $\alpha = c_v^{(e)}/c_v$ and for a fixed value of δ: theoretical results arising from Eq. (6.39) in the case of $f = f(\varrho_e)$

higher α, the higher η_{opt}. We also note that in Fig. 6.2 the value $\eta_{\mathrm{opt}}/\eta_c = 0.13$, attained when $\beta_1 = 1$, corresponds to the case of a single-temperature model, namely, $T_e = T_p \equiv T$.

In Fig. 6.3, instead, we plot the behavior of the ratio $\eta_{\mathrm{opt}}/\eta_c$ as a function of δ, for three different values of the non-dimensional parameter α, i.e., $\alpha = 0.05$ and $\alpha = 0.75$, and when $\beta_1 = 0.5$. In obtaining the results of Fig. 6.3 we still assumed that $\Lambda_p = \Lambda_e$, and $Z_{\mathrm{eff}}T = 1$. Along with previous observations about the role of the gradient of the electric charge, from Fig. 6.3 we infer that the smaller δ, the higher the thermoelectric performances. Figure 6.3 still allows to analyze the role of the different specific heats. In particular, we have that the smaller α, the higher η_{opt}. We note that these behaviors are in agreement with those of Fig. 6.2, since we are assuming that $\beta_1 < 1$. We would have the opposite behaviors if $\beta_1 > 1$.

For practical applications, these results suggest that, in order to enhance the efficiency of thermoelectric energy conversion, it would be useful to have:

- In a single-component material, the charge distribution as much homogeneous as possible, in such a way that $\bar{f} \to 0$ in Eq. (6.35).
- In a graded thermoelectric material, where the inhomogeneity in the electric charge arises naturally, the gradient of function f defined in Eq. (6.32) as small as possible.
- The physical parameter $\dfrac{\beta_1}{\beta_2} = \dfrac{T_e}{T_p}$ as high as possible.

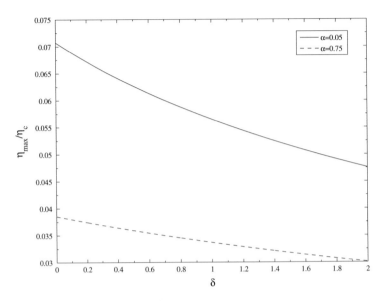

Fig. 6.3 Behavior of $\eta_{\max}/\eta_c$ versus δ for two different values of the non-dimensional parameter $\alpha = c_v^{(e)}/c_v$ and for $\beta_1 = 0.5$: theoretical results arising from Eq. (6.39) in the case of $f = f(\varrho_e)$

- The electron specific heat $c_v^{(e)}$ as high (respectively, small) as possible, and the phonon specific heat $c_v^{(p)}$ as small (respectively, high) as possible, whenever $\dfrac{\beta_1}{\beta_2} > 1$ (respectively, $\dfrac{\beta_1}{\beta_2} < 1$).

6.2 One-Temperature Models: Effects of Non-local and Non-linear Breakings of Onsager Symmetry

On microscopic grounds both electrons and phonons may be viewed as a free-particle gas in a box [1, 11, 35, 56–58]. In nonequilibrium statistical mechanics, the statistical behavior of a thermodynamic system far from its thermodynamic equilibrium is described through the Boltzmann transport equation (BTE), which in the relaxation-time approximation reads:

$$\frac{\partial f}{\partial t} + \mathbf{v} \cdot \nabla_{\mathbf{r}} f + \frac{\mathbf{F}}{m} \cdot \nabla_{\mathbf{v}} f = -\frac{f - f_0}{\tau} \qquad (6.40)$$

wherein the subscripts $\mathbf{r}$ and $\mathbf{v}$ in the nabla operators represent the variables of the gradient, i.e., they are the position and the velocity, respectively. Moreover, in Eq. (6.40) $\mathbf{F}$ is an external force instantly acting on each particle, m is the particle mass, $f(\mathbf{r}, \mathbf{v}, t)$ is the unknown distribution function, f_0 represents the equilibrium distribution of the carrier, and $\tau\left(\mathbf{r}, \mathbf{k}\right)$ is their relaxation time, $\mathbf{k}$ being the wave vector of the particle. In the BTE, f_0 is given by the Bose-Einstein distribution function (BEdf) in the case of phonons, i.e.,

$$f_0 = \left(e^{\frac{\hbar v}{k_B T}} - 1 \right)^{-1}$$

wherein $\hbar = h/(2\pi)$ with h being the Planck constant, and v is the angular frequency. In the case of electrons, instead, f_0 is expressed by the Fermi-Dirac distribution function (FDdf), i.e.,

$$f_0 = \left(e^{\frac{\varepsilon_i - \mu_e}{k_B T}} + 1 \right)^{-1}$$

with ε_i being the energy of the single-particle state.

Indeed, it is possible to find several situations, as for example whenever the quantities $(\varepsilon_i - \mu_e)$ and $\hbar v$ are much larger than $k_B T$, in which one can ignore the ± 1 in the denominator of f_0, and the BEdf and the FDdf reduce to the Boltzmann distribution function [11]. In these cases, on intuitive grounds, the solution of the BTE both for phonons, and for electrons would lead to equations showing the same mathematical behavior.

Starting from these considerations and in accordance with the basic principles of EIT [34, 40], generalized transport equations to describe heat and electric transport with thermoelectric coupling can be introduced [14, 58, 59].

6.2.1   The Nonlocal Model

Incorporating explicitly the effects of the several mean-free paths is useful to study new strategies for the optimization of thermoelectric effects. This problem has been addressed from a microscopic point of view by solving the Boltzmann equation for phonons and electrons, for nanowires made of InSb, InAs, GaAs and InP. Indeed, in the framework of EIT, in [36] it is proven that Eqs. (6.3), for vanishing values of μ_e/z_e, can be generalized as

$$\tau_p \dot{\mathbf{q}}^{(p)} + \mathbf{q}^{(p)} = -\lambda_p \nabla T + \ell_p^2 \left(\nabla^2 \mathbf{q}^{(p)} + 2\nabla\nabla \cdot \mathbf{q}^{(p)} \right) \tag{6.41a}$$

$$\tau_e \dot{\mathbf{q}}^{(e)} + \mathbf{q}^{(e)} = -\left(\lambda_e + \epsilon \Pi \sigma_e \right) \nabla T + \ell_e^2 \left(\nabla^2 \mathbf{q}^{(e)} + 2\nabla\nabla \cdot \mathbf{q}^{(e)} \right) + \Pi \sigma_e \mathbf{E} \tag{6.41b}$$

$$\tau_e \dot{\mathbf{I}} + \mathbf{I} = \sigma_e \left(\mathbf{E} - \epsilon \nabla T \right) + \ell_e^2 \left(\nabla^2 \mathbf{I} + 2\nabla\nabla \cdot \mathbf{I} \right) \tag{6.41c}$$

where $(\tau_p; \ell_p)$ and $(\tau_e; \ell_e)$ are the relaxation time and the mean-free path of phonons and electrons [14, 28, 35, 58].

In the next we use a phonon-hydrodynamic approach and the thermodynamic model in Eqs. (6.41) to phenomenologically explore the size dependency of the figure-of-merit. Our aim is to bridge the gap between the much detailed microscopic approaches (i.e., kinetic theory or numerical simulations) and the classical nonequilibrium-thermodynamic approaches [17, 23, 44, 45] lacking the explicit presence of the mean-free path.

For the sake of illustration, we will consider cylindrical nanowires, with a longitudinal length L and with a radius R of the transversal section.

6.2.1.1 Cylindrical Nanowires with $\ell_e < R < \ell_p$

We start to focus our attention on nanowire which has a transversal radius R such that $\ell_e < R < \ell_p$. This is a realistic assumption in Si or Bi_2Te_3 nanowires, for example, wherein the phonon mean-free path is one order of magnitude larger than the electron one [47, 54].

In this situation, for the electrons one has the usual resistive regime whereas the phonons undergo the hydrodynamic regime. In the resistive regime the second-order spatial derivatives of the flux can be neglected in the evolution equation of the heat flux, and the main term is the flux itself. In the hydrodynamic regime, instead, the spatial derivatives of the heat flux play the main role, as we observed in Chap. 3. Therefore, Eqs. (6.41) become [58]

$$\nabla^2 \mathbf{q}^{(p)} = \frac{\lambda_p}{\ell_p^2} \nabla T \tag{6.42a}$$

$$\mathbf{q}^{(e)} = -(\lambda_e + \epsilon \Pi \sigma_e) \nabla T + \Pi \sigma_e \mathbf{E} \tag{6.42b}$$

$$\mathbf{I} = \sigma_e (\mathbf{E} - \epsilon \nabla T). \tag{6.42c}$$

Under the assumption that ∇T and $\mathbf{E}$ are homogeneous across and along the nanowire, and with a vanishing phonon heat flux at the walls (i.e., if $q^{(p)}(R) = 0$), from the system (6.42) it follows that in each transversal section the phonon heat flux has a parabolic profile, whereas the electron heat flux and the current density are constant, namely,

$$q^{(p)}(r) = \lambda_p \frac{\Delta T}{L} \frac{(R^2 - r^2)}{4\ell_p^2} \tag{6.43a}$$

$$q^{(e)} = q^{(e,\Delta T)} + q^{(e,E)} = (\lambda_e + \epsilon \Pi \sigma_e) \frac{\Delta T}{L} + \Pi \sigma_e E \tag{6.43b}$$

$$I = I^{(\Delta T)} + I^{(E)} = \sigma_e \left(\epsilon \frac{\Delta T}{L} + E \right) \tag{6.43c}$$

wherein we have explicitly expressed the two different contributions to the electron heat flux $q^{(e)}$: that due to the temperature gradient (i.e., $q^{(e,\Delta T)}$), and that due to the electric field (i.e., $q^{(e,E)}$). Analogously, in Eq. (6.43c) we have expressed the two different contributions to the electric-current density I, namely, $I^{(\Delta T)}$ due to the temperature gradient, and $I^{(E)}$ due to the presence of an electric field.

Equations (6.43) allow to define the following effective transport coefficients

$$(\lambda_e + \epsilon \Pi \sigma_e)^{\text{eff}} = \frac{L}{\Delta T} \left[\frac{Q_{\text{tot}}^{(e,\Delta T)}}{\pi R^2} \right] = \frac{L}{\Delta T} \left[\frac{\int_0^R 2\pi r q^{(e,\Delta T)} dr}{\pi R^2} \right] = \lambda_e + \epsilon \Pi \sigma_e$$

(6.44a)

$$(\Pi \sigma_e)^{\text{eff}} = \frac{1}{E} \left[\frac{Q_{\text{tot}}^{(e,E)}}{\pi R^2} \right] = \frac{1}{E} \left[\frac{\int_0^R 2\pi r q^{(e,E)} dr}{\pi R^2} \right] = \Pi \sigma_e$$

(6.44b)

$$\lambda_p^{\text{eff}} = \frac{L}{\Delta T} \left[\frac{Q_{\text{tot}}^{(p)}}{\pi R^2} \right] = \frac{L}{\Delta T} \left[\frac{\int_0^R 2\pi r q^{(p)}(r) dr}{\pi R^2} \right] = \frac{\lambda_p}{8 \, \text{Kn}_p^2},$$

(6.44c)

$$(\epsilon \sigma_e)^{\text{eff}} = \frac{L}{\Delta T} \left[\frac{J_{\text{tot}}^{(\Delta T)}}{\pi R^2} \right] = \frac{L}{\Delta T} \left[\frac{\int_0^R 2\pi r I^{(\Delta T)} dr}{\pi R^2} \right] = \epsilon \sigma_e$$

(6.44d)

$$\sigma_e^{\text{eff}} = \frac{1}{E} \left[\frac{J_{\text{tot}}^{(E)}}{\pi R^2} \right] = \frac{1}{E} \left[\frac{\int_0^R 2\pi r I^{(E)} dr}{\pi R^2} \right] = \sigma_e$$

(6.44e)

wherein $\text{Kn}_p = \ell_p / R$ is the phonon Knudsen number.

The theoretical predictions (6.44) point out that only the phonon contribution to the effective thermal conductivity λ_p shows the influence of the radius when $\ell_e < R < \ell_p$. The material functions λ_e^{eff}, σ_e^{eff}, Π^{eff}, and ϵ^{eff}, instead, take the bulk values. Therefore, in such a situation, a bigger value of Z with respect to the bulk situation may be only obtained by a reduction in λ_p.

When Eqs. (6.44) are introduced into Eq. (6.1), one obtains the following effective figure-of-merit

$$Z_{\text{eff}} = \frac{(\epsilon^{\text{eff}})^2 \sigma_e^{\text{eff}}}{\lambda_e^{\text{eff}} + \lambda_p^{\text{eff}}} = \frac{\epsilon^2 \sigma_e}{\lambda_e + \dfrac{\lambda_p}{8 \, \text{Kn}_p^2}}.$$

(6.45)

Table 6.1 Value of the material functions for a p-doped sample of Bi_2Te_3 at 300 K [25, 26]

λ_p (W/mK)	λ_e (W/mK)	ℓ_p (nm)	ℓ_e (nm)	c_v (J/m^{3}K)	σ_e (1/Ωm)	ϵ (V / °C)
1.6	2.4	3.0	0.9	1.2×10^6	2.0×10^5	2.0×10^{-4}

   For increasing values of the phonon Knudsen number Kn_p, from Eqs. (6.44c) it follows that λ_p^{eff} decreases quadratically, so that from Eq. (6.45) we infer that Z_{eff} increases with respect to its bulk value and tends quadratically to the limit value $\epsilon^2 \sigma_e / \lambda_e$. A similar behavior for the figure-of-merit as a function of the transversal radius of the nanowire may be found in Refs. [25, 26] in the case of Bi_2Te_3 nanowires (refer to Table 6.1 for the values of the corresponding material functions at room temperature).

   If, instead, along with the observations of Chap. 3 we take into account the wall contribution by assuming again that the phonon heat flux is $q^{(p)}(r) = q_b^{(p)}(r) + q_w^{(p)}$ [58], with the wall contribution given by Eq. (3.8a), then Z_{eff} changes in [58]

$$Z_{\text{eff}} = \frac{\epsilon^2 \sigma_e}{\lambda_e + \dfrac{\lambda_p}{8\,Kn_p^2}\left(1 + 4C_p\,Kn_p\right)} \qquad (6.46)$$

reducing to

$$Z_{\text{eff}} = \frac{\epsilon^2 \sigma_e}{\lambda_e + \lambda_p\left(\dfrac{C_p}{2\,Kn_p}\right)} \qquad (6.47)$$

for increasing values of Kn_p. As observed from Eq. (6.47), the figure-of-merit depends now linearly on the reciprocal value of Kn_p. For increasing values of Kn_p, Z_{eff} still increases and tends to the limit value $Z_{\text{lim}} = \epsilon^2 \sigma_e / \lambda_e$, but with a linear behavior, as it can be seen also from Fig. 6.4, wherein the behavior of $Z_{\text{eff}}T$, as a function of the ratio Kn_p, is plotted for different values of the numerical coefficient C_p in the case of a nanosample made of a p-doped Bi_2Te_3 at 300 K. Figure 6.4 also enlightens the crucial role of the coefficient C_p, describing the phonon-wall collisions. If the walls are rough [27], C_p will be small and the predicted Z_{eff} will exhibit an enhancement, with respect to the theoretical prediction (6.45) ($C_p = 0$ in Fig. 6.4). The values of the material functions are taken from Hicks and Dresselhaus [25, 26] (refer to Table 6.1 for these values at room temperature) in the case of Bi_2Te_3 nanowires, which are often used in thermoelectric applications [47, 54]. It is worth observing that the ratio Kn_p can be modified by varying either R or the temperature, since the mean-free path depends on the temperature.

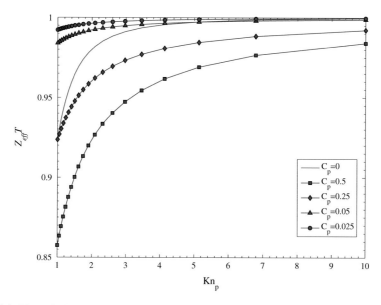

Fig. 6.4 Theoretical behavior of $Z_{\text{eff}}T$ in a p-doped Bi_2Te_3 nanosample at $300\,\text{K}$ as a function of the ratio Kn_p: comparison between Eq. (6.45) ($C_p = 0$) and Eq. (6.46) ($C_p \neq 0$) in the case of $\ell_e < R < \ell_p$. Different values for C_p have been taken into account, in order to explore the influence of the phonon-wall collisions on Z_{eff}. Refer to Table 6.1 for the values of the material functions

6.2.1.2   Cylindrical Nanowires with $\ell_e < \ell_p < R$

If one proceeds to a more severe miniaturization, R may also become smaller than the electron mean-free path (that is, $R < \ell_e, \ell_p$), so that both the heat carriers, and the current ones undergo the hydrodynamic regime. In this case, Eqs. (6.41) reduce to [58]

$$\nabla^2 \mathbf{q}^{(p)} = \frac{\lambda_p}{\ell_p^2} \nabla T \tag{6.48a}$$

$$\nabla^2 \mathbf{q}^{(e)} = \frac{1}{\ell_e^2} \left[(\lambda_e + \epsilon \Pi \sigma_e) \nabla T - \Pi \sigma_e \mathbf{E} \right] - 2 \nabla (\mathbf{E} \cdot \mathbf{I}) \tag{6.48b}$$

$$\nabla^2 \mathbf{I} = \frac{\sigma_e}{\ell_e^2} (\epsilon \nabla T - \mathbf{E}) \tag{6.48c}$$

which yield the following parabolic profiles in the bulk under the hypotheses that $\mathbf{E}$, $\mathbf{I}$ and ∇T are first-order quantities, and that second-order quantities are negligible [58]:

$$q_b^{(p)}(r) = \lambda_p \frac{\Delta T}{L} \frac{\left(R^2 - r^2 \right)}{4\ell_p^2} \tag{6.49a}$$

$$q_{\mathrm{b}}^{(e)}(r) = q_{\mathrm{b}}^{(e,\Delta T)}(r) + q_{\mathrm{b}}^{(e,\mathrm{E})}(r) = \left[(\lambda_e + \epsilon \Pi \sigma_e) \frac{\Delta T}{L} + \Pi \sigma_e \mathrm{E} \right] \frac{(R^2 - r^2)}{4\ell_e^2} \tag{6.49b}$$

$$I_{\mathrm{b}}(r) = I_{\mathrm{b}}^{(\Delta T)}(r) + I_{\mathrm{b}}^{(\mathrm{E})}(r) = \sigma_e \left(\epsilon \frac{\Delta T}{L} + \mathrm{E} \right) \frac{(R^2 - r^2)}{4\ell_e^2}. \tag{6.49c}$$

To account for the interactions between the different carriers and the walls [58], we may assume that

$$q^{(e)}(r) = q_{\mathrm{b}}^{(e)}(r) + q_{\mathrm{w}}^{(e)} \tag{6.50a}$$

$$q^{(p)}(r) = q_{\mathrm{b}}^{(p)}(r) + q_{\mathrm{w}}^{(p)} \tag{6.50b}$$

$$I(r) = I_{\mathrm{b}}(r) + I_{\mathrm{w}} \tag{6.50c}$$

with the wall contributions given by

$$q_{\mathrm{w}}^{(e)} = C_e \ell_e \left| \frac{\partial q_{\mathrm{b}}^{(e)}}{\partial r} \right|_{r=R} \tag{6.51a}$$

$$q_{\mathrm{w}}^{(p)} = C_p \ell_p \left| \frac{\partial q_{\mathrm{b}}^{(p)}}{\partial r} \right|_{r=R} \tag{6.51b}$$

$$I_{\mathrm{w}} = C_e \ell_e \left| \frac{\partial I_{\mathrm{b}}}{\partial r} \right|_{r=R} \tag{6.51c}$$

wherein C_e plays the same role of C_p, namely, it is a non-negative numerical coefficient characterizing the specular and diffusive reflections of the electrons hitting the walls. The coupling of Eqs. (6.44) with Eqs. (6.50) and (6.51) turns out that the different effective material functions are now given by

$$\lambda_e^{\mathrm{eff}} = \frac{\lambda_e}{8 \, \mathrm{Kn}_e^2} \left(1 + 4 C_e \, \mathrm{Kn}_e \right) \tag{6.52a}$$

$$\lambda_p^{\mathrm{eff}} = \frac{\lambda_p}{8 \, \mathrm{Kn}_p^2} \left(1 + 4 C_p \, \mathrm{Kn}_p \right) \tag{6.52b}$$

$$\sigma_e^{\mathrm{eff}} = \frac{\sigma_e}{8 \, \mathrm{Kn}_e^2} \left(1 + 4 C_e \, \mathrm{Kn}_e \right) \tag{6.52c}$$

$$\Pi^{\mathrm{eff}} = \Pi \tag{6.52d}$$

$$\epsilon^{\mathrm{eff}} = \epsilon \tag{6.52e}$$

with $\mathrm{Kn}_e = \ell_e/R$. Equations (6.52) lead to the following figure-of-merit [58]

$$Z_{\text{eff}} = \frac{\epsilon^2 \sigma_e \left(1 + 4 C_e \, \mathrm{Kn}_e\right)}{\lambda_e \left(1 + 4 C_e \, \mathrm{Kn}_e\right) + \lambda_p \dfrac{\ell_e^2}{\ell_p^2} \left(1 + 4 C_p \, \mathrm{Kn}_p\right)}. \tag{6.53}$$

For very small R, the Knudsen numbers Kn_p and Kn_e become dominant, and Eq. (6.53) reduces to

$$Z_{\text{eff}} = \frac{\epsilon^2 \sigma_e}{\lambda_e + \lambda_p \left(\dfrac{C_p}{C_e}\right) \left(\dfrac{\ell_e}{\ell_p}\right)}. \tag{6.54}$$

In this case an enhancement of Z_{eff} could be achieved by controlling the roughness of the walls, which determines the relative values of C_p and C_e. In particular, Eq. (6.54) suggests that Z_{eff} can be changed by modeling the walls of the nanowire in such a way that the number of reflected electrons is different from the number of reflected phonons, i.e., varying the ratio C_p/C_e. This can be also observed from Fig. 6.5, which plots the behavior of $Z_{\text{eff}}T$ in a p-doped nanosample made of $\mathrm{Bi}_2\mathrm{Te}_3$ at 300 K as a function of the ratio C_p/C_e. The analysis of Fig. 6.5 suggests that an enhancement of the figure-of-merit may be obtained when the number of

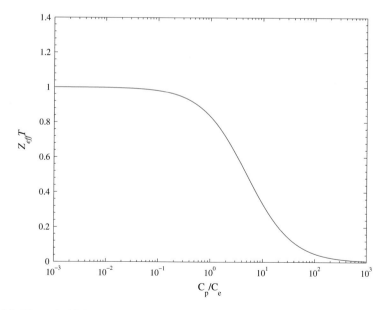

Fig. 6.5 Theoretical behavior of $Z_{\text{eff}}T$, arising from Eq. (6.54), in a p-doped $\mathrm{Bi}_2\mathrm{Te}_3$ nanosample at 300 K as a function of the ratio C_p/C_e in the case of $R < \ell_e, \ell_p$. The different values of the material functions are taken from Table 6.1. In figure the x-axis is in a logarithmic scale

reflected electrons is bigger than that of reflected phonons, i.e., the bigger C_e than C_p, the higher Z_{eff}.

6.2.2 The Crossed-Effects Nonlocal Model

Recently, nonlinear generalizations of the constitutive equations describing thermo-electric effects are being explored in a search of new possible strategies to improve the efficiency of thermoelectric energy conversion. In such generalizations, nonlinear coupling terms are incorporated into the equations, and microscopic expressions for their respective transport coefficients are looked for [9, 12, 52, 53, 60].

Indeed, in a more general setting, in EIT the model equations (6.41) can be also written as

$$\tau_p \dot{\mathbf{q}}^{(p)} + \mathbf{q}^{(p)} = -\lambda_p \nabla T + \nabla \cdot \mathbf{Q}^{(p)} \tag{6.55a}$$

$$\tau_e \dot{\mathbf{q}}^{(e)} + \mathbf{q}^{(e)} = -\lambda_e \nabla T + \nabla \cdot \mathbf{Q}^{(e)} + \Pi \mathbf{I} \tag{6.55b}$$

$$\tau_e \dot{\mathbf{I}} + \mathbf{I} = \sigma_e \left(\mathbf{E} - \epsilon \nabla T \right) + \nabla \cdot \mathbf{Q}^{(I)} \tag{6.55c}$$

for vanishing values of μ_e / z_e. In Eqs. (6.55) $\mathbf{Q}^{(p)}$, $\mathbf{Q}^{(e)}$ and $\mathbf{Q}^{(I)}$ stand for second-order tensors which, under suitable hypotheses in the state space, may be regarded, respectively, as the flux of $\mathbf{q}^{(p)}$, the flux of $\mathbf{q}^{(e)}$ and the flux of $\mathbf{I}$. The general form of Eqs. (6.55) makes them useful to study the consequences of the presence of these higher-order fluxes in the more general context of thermoelectricity. For example, let us assume the following constitutive equations [56]

$$\mathbf{Q}^{(p)} = \ell_p^2 \nabla \mathbf{q}^{(p)} \tag{6.56a}$$

$$\mathbf{Q}^{(e)} = \ell_e^2 \nabla \mathbf{q}^{(e)} + \alpha_e \mathbf{q}^{(e)} \mathbf{I} \tag{6.56b}$$

$$\mathbf{Q}^{(I)} = \ell_e^2 \nabla \mathbf{i} + \alpha_e' \mathbf{I} \mathbf{q}^{(e)} \tag{6.56c}$$

with α_e and α_e' being suitable coupling terms between the electron heat flux, the electric-current density and their first-order spatial derivatives [56]. A detailed microscopic analysis of these coefficients may be found in Ref. [56].

Equations (6.56) have the advantage of being simple and manageable, and point out the role of crossed effects in the simplest possible way. In fact, in steady states, by straightforward calculations the coupling of Eqs. (6.55) and (6.56) turns out the following generalization of Eq. (6.3) [56]

$$\mathbf{q} = -\lambda \nabla T + \widetilde{\Pi} \mathbf{I} \tag{6.57a}$$

$$\mathbf{I} = \widetilde{\sigma}_e \left(\mathbf{E} - \widetilde{\epsilon} \nabla T \right) \tag{6.57b}$$

wherein

$$\widetilde{\Pi} = \Pi\Pi^* = \Pi\left[1 + \frac{\alpha_e\sigma_e\mathbf{E}}{\Pi}\cdot(\mathbf{E} - \epsilon\nabla T)\right] \tag{6.58a}$$

$$\widetilde{\sigma}_e = \sigma_e\sigma_e^* = \sigma_e\left[1 + \alpha_e'\sigma_e\left(\mathbf{E}\cdot\mathbf{E} + \epsilon^2\nabla T\cdot\nabla T\right)\right] \tag{6.58b}$$

$$\widetilde{\epsilon} = \epsilon\epsilon^* = \epsilon\left[1 + \alpha_e'\sigma_e\left(\mathbf{E}\cdot\mathbf{E} - \epsilon^2\nabla T\cdot\nabla T\right)\right]. \tag{6.58c}$$

In the classical linear theory, the Onsager relations (OR) [44, 45], linking $\mathbf{q}$ and $\mathbf{I}$ to their respective thermodynamic forces ∇T^{-1} and $\mathbf{E}$, yield the relation $\Pi = \epsilon T$. In the present case, instead, by taking into account Eqs. (6.58a) and (6.58c) one obtains

$$\widetilde{\Pi} - \widetilde{\epsilon}T = a\mathbf{E}\cdot\mathbf{E} + b\nabla T\cdot\nabla T + c\mathbf{E}\cdot\nabla T \tag{6.59}$$

with

$$a = \left(\alpha_e - \alpha_e'\epsilon T\right)\sigma_e \tag{6.60a}$$

$$b = \alpha_e'\sigma_e\epsilon^3 T \tag{6.60b}$$

$$c = -\alpha_e\sigma_e\epsilon \tag{6.60c}$$

and a nonlinear breaking of the OR between the effective transport coefficient clearly ensues. Besides appealing from the theoretical point of view, these results are also interesting in practical applications. In particular, for a thermoelectric energy generator the maximum efficiency, as a function of the ratio x in Eq. (6.28), reads [56]

$$\eta_{max} = \eta_c\left(1 + \Psi_1\right)\left\{\frac{ZT + 2\left(1 - \Psi_2\right)\left[1 - \sqrt{1 + ZT\left(1 + \Psi_2\right)}\right]}{ZT}\right\} \tag{6.61}$$

wherein Ψ_1 and Ψ_2 are dimensionless parameters given as

$$\Psi_1 = \sigma_e\left\{\alpha_e'\left(\mathbf{E}\cdot\mathbf{E} - \epsilon^2\nabla T\cdot\nabla T\right) - \frac{\alpha_e\mathbf{E}}{\Pi}\cdot(\mathbf{E} - \nabla T)\right\} \tag{6.62a}$$

$$\Psi_2 = \sigma_e\left[2\alpha_e'\mathbf{E}\cdot\mathbf{E} + \frac{\alpha_e\mathbf{E}}{\Pi}\cdot(\mathbf{E} - \nabla T)\right] \tag{6.62b}$$

which vanish in the absence of coupling effects in Eqs. (6.56). Equation (6.61) shows that for a miniaturized system (the characteristic size of which is of the order of the mean-free path of the heat carriers), the maximum efficiency is not only a function of ZT, but it is also related to the degree of the Onsager symmetry (OS) breaking.

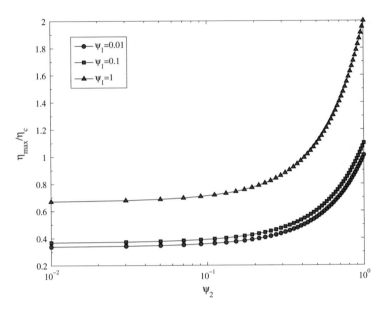

Fig. 6.6 Behavior of the ratio η_{max}/η_C as a function of Ψ_2 for different values of Ψ_1. For the sake of illustration we assumed $\Psi_1 = 0.01$, $\Psi_1 = 0.1$ and $\Psi_1 = 1$. Theoretical results arising from Eq. (6.61). In figure the x-axis is in a logarithmic scale

In Fig. 6.6 we plot the behavior of the ratio η_{max}/η_c as a function of Ψ_2 for increasing strength of Ψ_1 ($\Psi_1 = 0.01$, $\Psi_1 = 0.1$, $\Psi_1 = 1$). For the sake of computation, we assume that ψ_2 spans from 10^{-2} to 10^{-1}. Moreover, for the sake of illustration, we assume that the non-dimensional product $ZT = 3$, as for a Bi-doped n-type PbSeTe/PbTe quantum-dot superlattice sample at 500 K grown by molecular beam epitaxy [24].

As it can be seen, increasing values of Ψ_1 and/or of Ψ_2 lead to an enhancement of the thermoelectric efficiency with respect to the absence of nonlinear coupling terms, since in that case the ratio η_{max}/η_c would be equal to 0.33.

6.2.3   The Nonlinear Model

Since the early days in the study of thermoelectric phenomena it was seen that $\Pi = \epsilon T$. This relation was proved by Onsager [10, 44, 45] in a more general setting on reciprocity relations between coupling coefficients in coupled linear constitutive equations.

Although one may find many examples in which the OR comply with the experimental evidences [19, 23, 34], in some situations, especially in the nonlinear regime, the OR are no longer valid. As we previously observed, the problem of checking the validity of the OR goes beyond the pure theoretical interests, since these relations could have also consequences in practical applications. For example, in maximizing the efficiency of the thermoelectric energy conversion, we will show that when the OS holds, the maximum efficiency depends only on the so-called figure-of-merit, but a more general expression is required if such symmetry is broken.

To cope with nonlinear effects in thermoelectric applications, in Ref. [13] nonlinear thermoelectric equations have been introduced on a mesoscopic level to generalize Eqs. (6.3). Whenever terms proportional to $\nabla (\mu_e/z_e)$ are neglected, those equations read

$$\tau_p \dot{\mathbf{q}}^{(p)} + \mathbf{q}^{(p)} = -\lambda_p \nabla T + \frac{2\tau_p}{c_v T} \left(\nabla \mathbf{q}^{(p)} \cdot \mathbf{q}^{(p)} + \mathbf{q}^{(p)} \nabla \cdot \mathbf{q}^{(p)} \right) \tag{6.63a}$$

$$\tau_e \dot{\mathbf{q}}^{(e)} + \mathbf{q}^{(e)} = -(\lambda_e + \epsilon \Pi \sigma_e) \nabla T + \frac{2\tau_e}{c_v T} \left(\nabla \mathbf{q}^{(e)} \cdot \mathbf{q}^{(e)} + \mathbf{q}^{(e)} \nabla \cdot \mathbf{q}^{(e)} \right) + \Pi \sigma_e \mathbf{E} \tag{6.63b}$$

$$\tau_e \dot{\mathbf{I}} + \mathbf{I} = \sigma_e (\mathbf{E} - \epsilon \nabla T) + \frac{\tau_e}{\varrho_e} (\nabla \mathbf{I} \cdot \mathbf{I} + \mathbf{I} \nabla \cdot \mathbf{I}) \tag{6.63c}$$

wherein all the thermo-physical quantities are supposed to be constant, so that the nonlinearity is only due to the product of the partial heat fluxes and of the electrical current with their gradients. However, in the most general case the material functions are temperature dependent, and introduce a further nonlinearity in the system of equations. Here, such a situation is not considered, for the sake of simplicity.

In steady states, and whenever $\mathbf{q}$ and $\mathbf{I}$ may only vary along the longitudinal direction y of the system, Eqs. (6.63) reduce to [13]

$$q^{(p)} = -\lambda_p \nabla_y T \tag{6.64a}$$

$$q^{(e)} = -(\lambda_e + \epsilon \Pi \sigma_e) \nabla_y T + \left(\frac{8\tau_e}{c_v T} \right) EI q^{(e)} + \Pi \sigma_e E \tag{6.64b}$$

$$I = \sigma_e \left(E - \epsilon \nabla_y T \right) \tag{6.64c}$$

which allow to introduce the following effective Peltier coefficient [13]

$$\Pi^{\text{eff}} = \Pi \left[1 - \left(\frac{8\tau_e}{c_v T} \right) EI \right]^{-1} \tag{6.65}$$

whereas the other material functions coincide with their own bulk values (i.e., $\lambda^{\text{eff}} \equiv \lambda$, $\epsilon^{\text{eff}} \equiv \epsilon$ and $\sigma_e^{\text{eff}} \equiv \sigma_e$). Equation (6.65) shows a nonclassical part of the effective Peltier coefficient related to nonlinear effects following by the coupling between $\mathbf{q}^{(e)}$ and $\mathbf{I}$ both in Eq. (6.6b), and in Eq. (6.64b). Equation (6.65) also allows to point out the influence of nonlinear effects on the breaking of the OS at nanoscale.

In fact, if one assumes that $\Pi - \epsilon T = 0$ (i.e., OR holding for the bulk coefficients), then from that relation it follows

$$\Pi^{\mathrm{eff}} - \epsilon^{\mathrm{eff}} T = \left(\frac{8\tau_e \epsilon}{c_v}\right) Ei \left[1 - \left(\frac{8\tau_e}{c_v T}\right) EI\right]^{-1} \approx \left(\frac{8\tau_e \epsilon}{c_v}\right) EI \qquad (6.66)$$

namely, the second Kelvin relation breaks down for the effective coefficients. In particular, in this case, the following nondimensional parameter

$$\xi^* = \frac{\Pi^{\mathrm{eff}}}{\epsilon^{\mathrm{eff}} T} \approx 1 + \left(\frac{8\tau_e}{c_v T}\right) EI \qquad (6.67)$$

accounts for the breaking above. Whenever $\xi^* = 1$ (i.e., when the term $\tau_e EI/(c_v T)$ vanishes), the usual OR are recovered. Thus, at a given temperature, the degree of violation of the OS depends on the material functions τ_e and c_v, as well as on the intensity of the electrical field and of the electrical current, namely, the higher the product EI, the higher the deviation $\Pi^{\mathrm{eff}} - \epsilon^{\mathrm{eff}} T$ from zero.

The aforementioned breaking of the second Kelvin relation at nanoscale also has some interesting consequences on the efficiency (6.4) of a thermoelectric energy generator. In fact, in such a case the maximum attainable $\eta_{\max}$ becomes

$$\eta_{\max} = \frac{\eta_c}{\xi^*} \left(\frac{ZT\xi^* + 2 - 2\sqrt{1 + ZT\xi^*}}{ZT\xi^*}\right) = \frac{\eta_c}{\xi^*} \left(\frac{\sqrt{ZT\xi^* + 1} - 1}{\sqrt{ZT\xi^* + 1} + 1}\right) \qquad (6.68)$$

showing that nonlinear effects may be used to reach higher values of η. In fact, for a given value of the figure-of-merit, from Eq. (6.68) we also infer that the smaller ξ^*, the higher $\eta_{\max}$. This means that the thermoelectric efficiency also depends on the degree of the breaking of the second Kelvin relation for the effective Seebeck and Peltier coefficients.

This result is illustrated in Fig. 6.7, wherein we plot the thermoelectric efficiency as a function of ξ^*, for two different values of ZT. It is worth noticing that the effects of the breaking of the second Kelvin relation on the efficiency are not negligible, and it is higher for larger values of Z. Thus, in order to enhance the thermoelectric efficiency, it would be desirable to act both on ZT and ξ^*, making the first as large as possible, and the latter as small as possible. This last task can be achieved for moderate values of the product EI and using materials for which the ratio $8\tau_e/(c_v T)$ is as small as possible, in such a way that the values of the effective and of the bulk Peltier coefficients are sensibly close. As in the case of the model equations (6.63) the nondimensional parameter ξ^* is bigger than 1, according with Fig. 6.7 the breaking of the OR arising from nonlinear effects reduces the maximum efficiency.

A result analogous to Eq. (6.68) was found in Ref. [4] under the assumption that the time-reversal symmetry breaks down. In that paper the authors wrote the Onsager matrix assuming that the material functions depend on the external

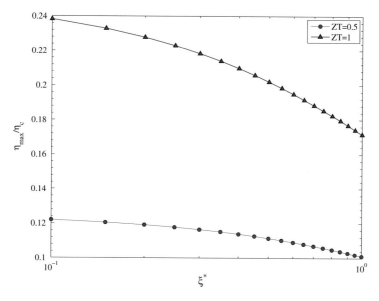

Fig. 6.7 Behavior of the ratio between the maximum thermoelectric efficiency and the Carnot efficiency (i.e., $\eta_{\max}/\eta_c$) versus the nondimensional ratio $\xi^* = \Pi^{\text{eff}}/\epsilon^{\text{eff}}T$, for two fixed values of ZT (i.e., $ZT = 0.5$ and $ZT = 1$). The results arise from Eq. (6.68), wherein the nondimensional parameter ξ^* accounts for the degree of the breaking of the second Kelvin relation. The case $\xi^* = 1$ corresponds to situation where the OS holds

magnetic field **B**. They assumed that $\epsilon(\mathbf{B}) \neq \epsilon(-\mathbf{B})$, in contrast to the equality required by OS [10], whereas $\lambda(\mathbf{B}) = \lambda(-\mathbf{B})$ and $\sigma_e(\mathbf{B}) = \sigma_e(-\mathbf{B})$. The maximum value of the thermoelectric efficiency was calculated as a function of the nondimensional ratio

$$x = \frac{\epsilon(\mathbf{B})}{\epsilon(-\mathbf{B})}$$

and it was found to be

$$\eta_{\max} = \eta_c \xi \frac{\sqrt{Z^*T + 1} - 1}{\sqrt{Z^*T + 1} + 1} \tag{6.69}$$

where

$$Z^* = \frac{\sigma_e(\mathbf{B})\,\epsilon(\mathbf{B})\,\epsilon(-\mathbf{B})}{\lambda(\mathbf{B})}$$

is the effective figure-of-merit. It is worth noticing that Eqs. (6.68) and (6.69) coincide if one sets $\xi = 1/\xi^*$ and $Z^* = Z\xi^*$. However, our analysis is more general, because it does not assume that the breaking of the OR is related to the magnetic field, but to nonlinear effects.

References

1. Alvarez, F.X., Jou, D., Sellitto, A.: Phonon hydrodynamics and phonon-boundary scattering in nanosystems. J. Appl. Phys. **105**, 014317 (5 pp.) (2009)
2. Balkan, N.: Hot Electrons in Semiconductors - Physics and Devices. Oxford University Press, New York (1998)
3. Benedict, L.X., Louie, S.G., Cohen, M.L.: Heat capacity of carbon nanotubes. Solid State Commun. **100**, 177–180 (1996)
4. Benenti, G., Saito, K., Casati, G.: Thermodynamic bounds on efficiency for systems with broken time-reversal symmetry. Phys. Rev. Lett. **106**, 230602 (4 pp.) (2011)
5. Berciaud, S., Han, M.Y., Mak, K.F., Brus, L.E., Kim, P., Heinz, T.F.: Electron and optical phonon temperatures in electrically biased graphene. Phys. Rev. Lett. **104**, 227401 (4 pp.) (2010)
6. Birman, V., Byrd, L.W.: Modeling and analysis of functionally graded materials and structures. Appl. Mech. Rev. **60**, 197–216 (2007)
7. Boukai, A.I., Bunimovich, Y., Tahir-Kheli, J., Yu, J.-K., Goddard-III, W.A., Heath, J.R.: Silicon nanowires as efficient thermoelectric materials. Nature **451**, 168–171 (2008)
8. Bulusu, A., Walker, D.: Review of electronic transport models for thermoelectric materials. Superlattices Microstruct. **44**, 1–36 (2008)
9. Büttiker, M.: Capacitance, admittance, and rectification properties of small conductors. J. Phys. Condens. Matter **55**, 9361–9378 (1993)
10. Casimir, H.B.G.: On Onsager's principle of microscopic reversibility. Rev. Mod. Phys. **17**, 343–350 (1945)
11. Chen, G.: Nanoscale Energy Transport and Conversion - A Parallel Treatment of Electrons, Molecules, Phonons, and Photons. Oxford University Press, Oxford (2005)
12. Christen, T., Büttiker, M.: Gauge-invariant nonlinear electric transport in mesoscopic conductors. Europhys. Lett. **35**, 523–528 (1996)
13. Cimmelli, V.A., Sellitto, A., Jou, D.: A nonlinear thermodynamic model for a breaking of the Onsager symmetry and efficiency of thermoelectric conversion in nanowires. Proc. R. Soc. A **470**, 20140265 (14 pp.) (2014)
14. Cimmelli, V.A., Carlomagno, I., Sellitto, A.: Non-fourier heat transfer with phonons and electrons in a circular thin layer surrounding a hot nanodevice. Entropy **17**, 5157–5170 (2015)
15. Coleman, B.D., Dill, E.H.: Thermodynamic restrictions on the constitutive equations of electromagnetic theory. Z. Angew. Math. Phys. **22**, 691–702 (1971)
16. Crommie, M.F., Zettl, A.: Thermal conductivity of single-crystal Bi-Sr-Ca-Cu-O. Phys. Rev. B **41**, 10978–10982 (1990)
17. de Groot, S.R., Mazur, P.: Nonequilibrium Thermodynamics. North-Holland, Amsterdam (1962)
18. Dresselhaus, M.S., Chen, G., Tang, M.Y., Yang, R.G., Lee, H., Wang, D.Z., Ren, Z.F., Fleurial, J.-P., Gogna, P.: New directions for low-dimensional thermoelectric materials. Adv. Mater. **19**, 1043–1053 (2007)
19. Dufty, J.W., Rubí, J.M.: Generalized Onsager symmetry. Phys. Rev. A **36**, 222–225 (1987)
20. Gilbert, G.T.: Positive definite matrices and Sylvester's criterion. Am. Math. Mon. **08**, 44–46 (1991)
21. Goodson, K.E., Flik, M.I.: Electron and phonon thermal conduction in epitaxial high-T_c superconducting films. J. Heat Transf. Trans. ASME **115**, 17–25 (1993)
22. Gouin, H., Ruggeri, T.: Identification of an average temperature and a dynamical pressure in a multitemperature mixture of fluids. Phys. Rev. E **78**, 016303 (7 pp.) (2008)
23. Gyarmati, I.: Nonequilibrium Thermodynamics. Springer, Berlin (1970)
24. Harman, T.C., Walsh, M.P., La Forge, B.E., Turner, G.W.: Nanostructured thermoelectric materials. J. Electron. Mater. **34**, L19–L22 (2005)
25. Hicks, L.D., Dresselhaus, M.S.: Effect of quantum-well structures on the thermoelectric figure of merit. Phys. Rev. B **47**, 12727–12731 (1993)

26. Hicks, L.D., Dresselhaus, M.S.: Thermoelectric figure of merit of a one-dimensional conductor. Phys. Rev. B **47**, 16631–16634 (1993)
27. Hochbaum, A.I., Chen, R., Delgado, R.D., Liang, W., Garnett, E.C., Najarian, M., Majumdar, A., Yang, P.: Enhanced thermoelectric performance of rough silicon nanowires. Nature **451**, 163–167 (2008)
28. Hopkins, P.E., Norris, P.M., Phinney, L.M., Policastro, S.A., Kelly, R.G.: Thermal conductivity in nanoporous gold films during electron-phonon nonequilibrium. J. Nanomater. **2008**, 418050 (7 pp.) (2008)
29. Humphrey, T.E., Linke, H.: Reversible thermoelectric nanomaterials. Phys. Rev. Lett. **94**, 096601 (4 pp.) (2005)
30. Humphrey, T.E., Newbury, R., Taylor, R.P., Linke, H.: Reversible quantum Brownian heat engines for electrons. Phys. Rev. Lett. **89**, 116801 (4 pp.) (2002)
31. Iwanaga, S., Toberer, E.S., La Londe, A., Snyder, G.J.: A high temperature apparatus for measurement of the Seebeck coefficient. Rev. Sci. Instrum. **82**, 063905 (2011)
32. Jeffrey Snyder, G., Toberer, E.S.: Complex thermoelectric materials. Nat. Mater. **7**, 105–114 (2008)
33. Joshi, G., et al.: Enhanced thermoelectric figure-of-merit in nanostructured p-type silicon germanium bulk alloys. Nano Lett. **8**, 4670–4674 (2008)
34. Jou, D., Casas-Vázquez, J., Lebon, G.: Extended Irreversible Thermodynamics, 4th revised edn. Springer, Berlin (2010)
35. Jou, D., Cimmelli, V.A., Sellitto, A.: Nonlocal heat transport with phonons and electrons: application to metallic nanowires. Int. J. Heat Mass Transf. **55**, 2338–2344 (2012)
36. Jou, D., Sellitto, A., Cimmelli, V.A.: Multi-temperature mixture of phonons and electrons and nonlocal thermoelectric transport in thin layers. Int. J. Heat Mass Transf. **71**, 459–468 (2014)
37. Koechlin, F., Bonin, B.: Parametrisation of the Niobium thermal conductivity in the superconducting state. In: Bonin, B. (ed.) Proceedings of the 1995 Workshop on RF Superconductivity, Gif-sur-Yvette, France, pp. 665–669. Gordon and Breach, New York (1996)
38. Kuznetsov, V.L.: Functionally graded materials for thermoelectric applications. In: Rowe, D.M. (ed.) Thermoelectrics Handbook: Macro to Nano – Sec. 38. CRC Press, Boca Raton (2005)
39. Kuznetsov, V.L., Kuznetsova, L.A., Kaliazin, A.E., Rowe, D.M.: High performance functionally graded and segmented Bi_2Te_3-based materials for thermoelectric power generation. J. Mater. Sci. **37**, 2893–2897 (2002)
40. Lebon, G., Jou, D., Casas-Vázquez, J.: Understanding Nonequilibrium Thermodynamics. Springer, Berlin (2008)
41. Lin, Z., Zhigilei, L.V., Celli, V.: Electron-phonon coupling and electron heat capacity of metals under conditions of strong electron-phonon nonequilibrium. Phys. Rev. B **77**, 075133 (17 pp.) (2008)
42. Müller, E., Drašar, Č., Schilz, J., Kaysser, W.A.: Functionally graded materials for sensor and energy applications. Mater. Sci. Eng. **A362**, 17–39 (2003)
43. Nakamura, D., Murata, M., Yamamoto, H., Hasegawa, Y., Komine, T.: Thermoelectric properties for single crystal bismuth nanowires using a mean free path limitation model. J. Appl. Phys. **110**, 053702 (6 pp.) (2011)
44. Onsager, L.: Reciprocal relations in irreversible processes I. Phys. Rev. **37**, 405–426 (1931)
45. Onsager, L.: Reciprocal relations in irreversible processes II. Phys. Rev. **38**, 2265–2279 (1931)
46. Pokharela, M., Zhaoa, H., Lukasa, K., Rena, Z., Opeila, C., Mihaila, B.: Phonon drag effect in nanocomposite $FeSb_2$. MRS Commun. **3**, 31–36 (2013)
47. Qui, B., Sun, L., Ruan, X.: Lattice thermal conductivity reduction in Bi_2Te_3 quantum wires with smooth and rough surfaces: a molecular dynamics study. Phys. Rev. B **83**, 035312 (7 pp.) (2011)
48. Rogolino, P., Sellitto, A., Cimmelli, V.A.: Influence of the electron and phonon temperature and of the electric-charge density on the optimal efficiency of thermoelectric nanowires. Mech. Res. Commun. **68**, 77–82 (2015)
49. Ruggeri, T.: Multi-temperature mixture of fluids. Theor. Appl. Mech. **36**, 207–238 (2009)

50. Ruggeri, T., Lou, J.: Heat conduction in multi-temperature mixtures of fluids: the role of the average temperature. Phys. Lett. A **373**, 3052–3055 (2009)
51. Ruggeri, T., Sugiyama, M.: Rational Extended Thermodynamics beyond the Monatomic Gas. Springer, Cham (2015)
52. Sánchez, D., Büttiker, M.: Magnetic-field asymmetry of nonlinear mesoscopic transport. Phys. Rev. Lett. **93**, 106802 (4 pp.) (2004)
53. Sánchez, D., López, R.: Scattering theory of nonlinear thermoelectric transport. Phys. Rev. Lett. **110**, 026804 (5 pp.) (2013)
54. Satterthwaite, C.B., Ure, R.W.J.: Electrical and thermal properties of Bi_2Te_3. Phys. Rev. **108**, 1164–1170 (1957)
55. Schreier, M., Kamra, A., Weiler, M., Xiao, J., Bauer, G.E.W., Gross, R., Goennenwein, S.T.B.: Magnon, phonon, and electron temperature profiles and the spin Seebeck effect in magnetic insulator/normal metal hybrid structures. Phys. Rev. B **88**, 094410 (2013)
56. Sellitto, A.: Crossed nonlocal effects and breakdown of the Onsager symmetry relation in a thermodynamic description of thermoelectricity. Phys. D **243**, 53–61 (2014)
57. Sellitto, A.: Frequency dependent figure-of-merit in cylindrical thermoelectric nanodevices. Phys. B **456**, 57–65 (2015)
58. Sellitto, A., Cimmelli, V.A., Jou, D.: Thermoelectric effects and size dependency of the figure-of-merit in cylindrical nanowires. Int. J. Heat Mass Transf. **57**, 109–116 (2013)
59. Sellitto, A., Cimmelli, V.A., Jou, D.: Influence of electron and phonon temperature on the efficiency of thermoelectric conversion. Int. J. Heat Mass Transf. **80**, 344–352 (2015)
60. Spivak, B., Zyuzin, A.: Signature of the electron-electron interaction in the magnetic-field dependence of nonlinear I-V characteristics in mesoscopic systems. Phys. Rev. Lett. **93**, 226801 (3 pp.) (2004)
61. Stojanovic, N., Maithripala, D.H.S., Berg, J.M., Holtz, M.: Thermal conductivity in metallic nanostructures at high temperature: electrons, phonons, and the Wiedemann-Franz law. Phys. Rev. B **82**, 075418 (9 pp.) (2010)
62. Straube, H., Wagner, J.-M., Breitenstein, O.: Measurement of the Peltier coefficient of semiconductors by lock-in thermography. Appl. Phys. Lett. **95**, 052107 (3 pp.) (2009)
63. Tzou, D.Y.: Macro- to Microscale Heat Transfer: The Lagging Behaviour, 2nd edn. Wiley, Chichester (2014)
64. Volz, S. (ed.): Thermal Nanosystems and Nanomaterials. Topics in Applied Physics. Springer, Berlin (2010)
65. Wang, X., Wang, Z. (eds.): Nanoscale Thermoelectrics. Springer, Berlin (2014)
66. Zemanski, M.W., Dittman, R.H.: Heat and Thermodynamics, 7th revised edn. MacGraw-Hill, New York (1997)
67. Zhang, Z.M.: Nano/Microscale Heat Transfer. McGraw-Hill, New York (2007)

Chapter 7
Perspectives

We close this book with some perspectives on open problems to be further analyzed. The first of them is, of course, the detailed microscopic understanding of the generalized transport equations, starting from phonon kinetic theory, and using several different techniques to work out approximate solutions of it. Several factors to be taken into account would be, for instance, the role of detailed phonon dispersion relations, the frequency and the temperature dependence of the phonon collision times, the role of the different effective phonon mean-free paths, the slip heat flow along the walls, and so on.

A second open problem is the detailed comparison with experiments for a variety of relevant materials, as Si, Ge, Si_xGe_{1-x}, Bi, Te, Bi_2Te_3, and so on, as well as for systems with different porosities and for graded materials. It would be useful to characterize in details the suitable material transport functions for such materials, in a wide range of temperatures. The role of phonon hydrodynamics in two-dimensional systems (graphene, silicon thin layers, microporous thin layers) should also be explored with more depth, because there are hints of their increased relevance as compared to three-dimensional systems. Radial heat transport from hot spots, and its application to temperature measurement (or to nanodevice refrigeration) is also a lively topic. Quantum confinement should also be considered at very low temperatures, or in very narrow channels. In such a case, the thermal phonon wavelength becomes comparable to the size of the system and the quantum discrete levels of energy become relevant, instead of the continuous approach.

It would be interesting to deal at length with time-dependent phenomena, for a wide range of frequencies. The use of thermal waves (not necessarily the very high-frequency ones) could be useful for characterizing some properties of the materials, as for instance the porosity, or the distribution of the radii of particles embedded in a heat conducting matrix.

Another topic worth of exploration would be the several devices necessary for the theoretical (or practical) development of phononics, namely, heat rectifiers and

© Springer International Publishing Switzerland 2016
A. Sellitto et al., *Mesoscopic Theories of Heat Transport in Nanosystems*,
SEMA SIMAI Springer Series 6, DOI 10.1007/978-3-319-27206-1_7

heat transistors, in order to enhance rectification and to achieve negative differential thermal conductivity.

More emphasis on thermoelectricity, and the analysis not only of heat conductivity, but also of electrical conductivity, Seebeck coefficient, Peltier coefficient, as well as their possible nonlinear contributions, should also be paid, in order to improve efficiencies of thermoelectric conversion.

In summary, the formulation of mesoscopic heat-transport equations, well characterized for technologically useful materials, and well connected with microscopic understanding of transport in such materials and with a generalized entropy, entropy flux and second law, seems to be truly one of the current most exciting frontiers in nonequilibrium thermodynamics.

Index